BIOTECHNOLOGY IN AGRICULTURE AND THE FOOD INDUSTRY

BIOTECHNOLOGY IN AGRICULTURE AND THE FOOD INDUSTRY

G. E. ZAIKOV

EDITOR

Nova Science Publishers, Inc.

New York

Production Coordinator: Donna Dennis
Coordinating Editor: Tatiana Shohov
Senior Production Editors: Susan Boriotti and Donna Dennis
Production Editors: Marius Andronie and Rusudan Razmadze
Office Manager: Annette Hellinger
Graphic Design: Magdalena Nuñez,
Editorial Production: Maya Columbus, Aloysha Klestov, Vladimir Klestov,
Matthew Kozlowski and Lorna Loperfido
Circulation: Ave Maria Gonzalez, Vera Popovic, Luis Aviles, Jonathan Boriotti, Alexandra Columbus, Raymond Davis, Cathy DeGregory, Melissa Diaz, Marlene Nuñez, Jeannie Pappas and Frankie Punger
Communications and Acquisitions: Serge P. Shohov

Library of Congress Cataloging-in-Publication Data
Biotechnology and agriculture and the food industry / G. E. Zaikov ... [et al.], editor.
p. cm.
Includes bibliographical references and index.
ISBN 1-59454-119-1 (hardcover)

400 Oser Ave, Suite 1600
Hauppauge, New York 11788-3619
Tele. 631-231-7269 Fax 631-231-8175
e-mail: Novascience@earthlink.net
Web Site: http://www.novapublishers.com

Printed in the United States of America

CONTENTS

PREFACE

G. E. Zaikov

*"If you perceive the Proper Way in the morning,
you can quietly die at night
with the sense of followed out duty"*
Confucius, China
(500 years A.C.)

"Knowledge is force"
Frances Bacon (1561-1626),
The lord-Chancellor,
Philosopher-materialist
England

We are not sure that anybody who is sane will willingly agree to die, but everybody wants to know the "Proper Way" in "Science" and someone in "Religion" (and stay alive whenever possible). We also tried to do this in the case of Biotechnology: state of the art and prospects for development.

This volume is dedicated to biotechnology in agriculture and the food industry.

We would like to say many thanks for Dr. Larisa P. Krylova who helped for us very much during preparation of this volume.

The Editor of volume will be happy to receive any kind of scientific comments about this book. We will use them in our research in future.

G.E. Zaikov,
Prof., Member of Directorium,
N.M. Emanuel Institute of Biochemical Physics,
Russian Academy of Sciences.
4 Kosygin st., Moscow- 119991, Russia
Fax 7 095 137 41 01
E-mail chembio@sky,chph.ras.ru

SECTION 1: BIOTECHNOLOGY AND AGRICULTURE

In: Biotechnology in Agriculture and the Food Industry ISBN 1-59454-119-1
Editor: G. E. Zaikov, pp. 3-9

Chapter 1

AGROBACTERIAL TRANSFORMATION OF PLANT CELLS: THEORETICAL AND BIOTECHNOLOGICAL ASPECTS

I. V. Volokhina, V. A. Velikov, A. V. Burmatov, A. A. Shirokov, O. S. Ermoshina and M. I. Chumakov*

Institute of Biochemistry and Physiology of Plants and Microorganisms
Russian Academy of Sciences,
13 prospekt Entuziastov, Saratov, 410049, Russia, fax: 7(8452)970383

ABSTRACT

We reveal the interaction of purified T-complex elements (ss-T-DNA, VirE2-protein) with flat black simulated membranes. Using the patch-clamp method, we detected a single voltage-dependent pore after contact of VirE2 protein with flat membranes. The life time for this type pore was 1.5 -7.0 sec. The calculated diameter of the pore was 2 nm. We developed the method of *Agrobacterium*-mediated plant genetic transformation without using bacteriostatic antibiotics. Agrobacterial mutants containing the *sac*B gene from *Bacillus subtilis* coding levansucrase were unable to grow on MS medium due to cell death resulting from osmotic shock

Key words: ss-T-DNA, VirE2 protein, membranes, transfer, protein pores, *Agrobacterium, Bacillus subtilis*, bacteriostatic antibiotic, transgenic plants

INTRODUCTION

Members of the genus *Agrobacterium* can produce undifferentiated tumors on a wide range of dicotyledonous plants after the agrobacterial T-DNA transfer into the plant genome.

* Corresponding author, e-mail address: chumakov@ibppm.sgu.ru

The mechanism of T-DNA transfer across host cell envelopes is still unclear. Under the influence of acetosyringone (AS) exudates by wounded plant cells, the T-DNA is liberated from the Ti-plasmid and is transferred as a nucleoprotein T-complex. The T-DNA, covalently linked to VirD2 (at the 5'end), is transported from bacterium to plant-cell nucleus. Both wild-type VirE2 and VirD2 proteins contain nuclear targeting sequences (NLS) that are likely to be involved in directing the transferred T strand to the plant nucleus. It was shown that *A.tumefaciens* separately transfers VirE2/VirE1-protein and the ss-T-DNA-VirD2 complex to the plant cell (Gelvin, 1998). The VirD2-T-DNA-VirE2-complex introduced into the plant cell, where it enters the nucleus and becomes integrated into the plant genome (Tinland, 1996). According to Dumas et al. (2001) VirE2 performs an unusually large number of different functions: (1) ss-DNA binding protein, (2) T-DNA protection from plant-nuclease degradation, (3) assistance in keeping the T-strand in an unfolded state, (4) formation of the VirE2 channel in lipid bilayers, (5) VirE2-protein transfer across the VirE2 channel, and (6) transport of the T-complex from the outer membrane to the nucleus.

It is believed, that short oligonucleotides can be transported through an artificial membrane via VirE2-protein-formed pores (Dumas et al., 2001). One aim of our work was to isolate ss-T-DNA, VirE2-protein and to investigate the interaction of T-complex elements with flat black simulated membranes.

On the other hand, the biological property of N-DNA transfer has a broad-scale practical application for transgenic plant production. In the process of plant transformation, it is necessary to prevent the multiplication of *Agrobacterium tumefaciens* on the medium for shoot regeneration from explants upon completion of the stage of their cultivation together with agrobacteria. Otherwise, regenerated plants will develop poorly because of the intensive bacterial growth and will finally die. The addition of antibiotics to the nutrient medium for plant cultivation (MS agar) is not quite reliable because of their degradation with time, and also unsafe for plant cells. We solved this problem by changing the transformation technique and using specially constructed agrobacterial strains.

RESULTS AND DISCUSSION

Purification of ssT-DNA

We isolated T-complex elements (ssT-DNA and VirE2 protein) from acetosyringone-induced agrobacterial cells. The T-DNA was purified by PCR amplification from pBI101.1-plasmid *Bam*H1 and *Sac*1-restriction fragments. Primers Pr1 5'-GCG TGG TAC CTC AGA AGA ACT CGT CAA G –3' and Pr2 5'- GTT TAC CCG CCA ATA TAT CCT GTC AAA CAC –3' were used. PCR fragment of dsT-DNA was eluted from agarose gel and used for ssT-DNA preparing.

Cloning of *virE2* into the Expression Vector pQE31

pTiC58 plasmid from *A.tumefaciens* C58 strain was used for the PCR synthesis of the *vir*E2 gene. After isolation of the *virE2*-containing fragment from gel, followed by the performance of a repeated PCR using this fragment as a matrix and the same primers. This used the following primers: 5'-GAC-GAT-GGA-TCC-GAA-GGC-CGA-AG-3' and 5'-AAAA-CTG-CAG-CTA-CAG-ACT-GTT-TAC-GGT-TGG- GCC-GC-3'.

PCR was performed at 50^0C with *Taq*-polymerase (Fermentas, Lithuania) for 30 cycles. A DNA-fragment corresponding to *vir*E2 gene with 1670 bp calculated size was obtained. The 1670 bp *vir*E2 containing fragment was eluted from the gel. The amplified PCR product was cleaved with *Bam*HI and *PstI* and inserted into plasmid pQE31 cleaved with the same restriction enzymes, resulting in pQE31- *vir*E2His6. The pQE31-*vir*E2 was checked for positive clones.

Overexpression and Purification of *virE2* Protein

pQE31 plasmid was introduced into *E.coli* XL-1 by electroporation. The transformed cells were grown in 500 mL of Luria-Bertani medium containing 100 µg/mL ampicillin at 37^0C. After 4-hr cultivation, expression was induced by addition of IPTG to a final concentration of 1 mM. Shaking was continued for 5 hr. The cells were then centrifuged at 6000 g for 30 min at 4^0C. The pellet was resuspended in 50 ml chilled lysis buffer A (50 мM NaH_2PO_4, pH 8.0, 300 мM NaCl, 5 мM imidazole, 1 mM PMSF, 0.5 mg/mL lysozyme). After 1 hr of incubation on ice, the mixture was centrifuged at 20000 g for 30 min at 4^0C. After centrifugation, the supernatant was applied to a Ni-NTA agarose column. Bound protein was eluted with buffer B (50 мM NaH_2PO_4, pH 8.0, 300 мM NaCl, 250 мM imidazole). VirE2 peak fractions were dialyzed overnight against buffer C (50 мM NaH_2PO_4, pH 8.0, 100 мM NaCl), centrifuged (20000 g for 10 min) to remove insoluble material, and stored at a concentration of 50 ng/mL. Purified His-tagged VirE2 protein is shown in Fig.1.

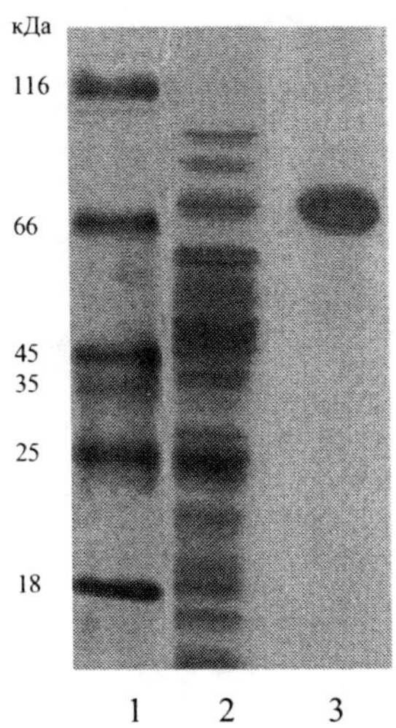

Fig. 1. SDS-PAAG of VirE2xHis-6 protein, 1. molecular weight markers (кDa), 2. cell extract of *E.coli* XL1-blue strain (pQE31-*vir*E2), 3. VirE2 protein purified using Ni-NTA-agarose

The VirE2 protein was identified using single-chain fragments variable (scFv) against VirE2 protein prepared by the phage display method and revealed with C-*myc* antibodies and labeling with anti-mouse antibodies with peroxidase (Fig. 2).

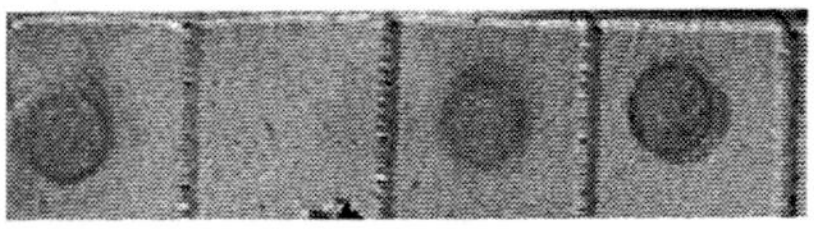

Fig.2. Immunodot analysis of VirE2 protein, 1. - crude cell extract of *A.tumefaciens* C58 after *vir*-gene induction with the acetosyringone; 2 - crude cell extract of *E.coli* XL-1 blue, 3 - crude cell extract of *E.coli* XL-1 blue after induction with 1 mM IPTG, carrying plasmid pQE31 with the *vir*E2-gene insertion; 4 - Ni-NTA purified VirE2 protein.

Obtainment of Miniantibodies Against *virE2* Protein

The mice library scFv constructed in *E.coli* strain TG1 was kindly provided by Dr. A. Laman (Pushchino Branch of the Institute of Bioorganic Chemistry, Russia). The repertoire of the library obtained is at least $5{\cdot}10^8$ recombinant clones. It was constructed in phagemid pHEN I in *E.coli* TG1; phage M13-K07 was used as a helper phage. The clones of phages producing antibodies to the agrobacterial virulence protein VirE2 have been isolated from this library by affinity selection, by using a standard procedure. These clones have been transfected into *E.coli* JS5 cells. We obtained 1.3-1.5 mg non-purified scFv from 1 L of *E.coli* cells. Soluble miniantibodies from strain JS5 were purified by affinity chromatography using Ni-NTA-agarose (Fig. 3). Some of the isolated scFv have been characterized immunochemically.

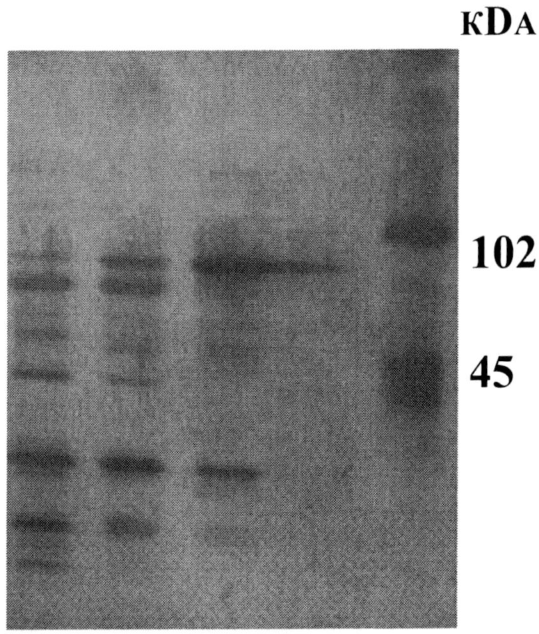

Fig.3. SDS-PAAG of scFv (miniantibodies) to VirE2 protein purified from *E.coli* strain TG1 extract using Ni-NTA-agarose. 1-4. scFv (miniantibodies) content in different fractions (pH 6.0, 5.0, 3.0, 2,5). 5. Molecular markers (κDa).

The Interaction of the *Agrobacterium* T-Complex Elements with Flat Membranes

Single-strend (ss) T-DNA of *Agrobacterium* in complex with VirD2 as well as VirE2 proteins can cross the cell membrane and the cell cytoplasm and can then be inserted in the host (plant, animal) chromosome. The mechanism of the membrane transfer is unknown. It was shown that short oligonucleotides can be transported through an artificial membrane via VirE2-protein-formed pores (Dumas et al., 2001).

We investigated the interaction of T-complex elements with flat lipid membranes. For preparation of membranes, we used phosphatidilcholine (50 - 100%), phosphatidil-ethanolamine (25 - 75%), and phosphatidilglycerol (1 - 10%). We revealed that 63.5% of phosphatidilcholine 30% of phosphatidiletanolamine, and 6.5% of phosphatidilglycerol are optimal for membrane stability.

The critical concentration of ss-binding protein from *E.coli*, VirE2-protein, ssT-DNA and T-complex (assembled) from *A.tumefaciens* was established. In particular, at a VirE2 concentration 16 ng/µl the membrane loses stability and breaks up. At ssT-DNA concentration (about 15000 b.p., the fragment of plasmid pBISNc) less then ~ 0,06 ng/µl the membrane behaves rather steadily; at concentration higher then 0.1 ng/ml, the stability is lost.

Using the patch-clamp method, we detected a single voltage-dependent pore after contact of VirE2 protein with flat membranes. The life time for this type pore was 1.5-7.0 sec. The calculated diameter of the pore was 2 nm.

Agrobacterium-Mediated Plant Transformation without Using Bacteriostatic Antibiotics

The method of preventing bacterial contamination of explants described here was used in the transformation of tobacco plants by the marker gene *nptII*. The agrobacterial cells were constructed sensitive to the main component of MS medium, sucrose, by transferring the *sacB* gene from *Bacillus subtilis* into the *A. tumefaciens* strain LBA4404. At the stage of co-cultivation of explants with agrobacteria, glucose was added to MS agar instead of sucrose in order to protect microbial cells from perishing until the genes are transferred into the plants. The standard MS agar was used at the induction stages of callus, shoot formation, and rooting of shoots.

First, the gene of levansucrase was introduced into the chromosome of the *A. tumefaciens* strain LBA4404 (pAL4404, pGA482) by using transposone mutagenesis with the transposone Tn5-B12S having the *sacB* gene. The mutants were unable to grow on the MS medium due to the cell death resulting from osmotic shock. Such mutants were obtained by mutagenesis of laboratory agrobacterial strains with the frequencies on the order of 10^{-6}. They were characterized by considerable instability. Because of the high frequency of reversions, it seemed impossible to completely prevent the multiplication of bacterial cells on MS agar. Therefore, the gene of levansucrase was later cloned into a plasmid compatible with the agrobacterial plasmids of the binary vector system to increase the stability of sucrose sensitivity.

On the basis of the broad-host range vector pSup106, recombinant constructions with the *sacB* gene were obtained (Fig.4).

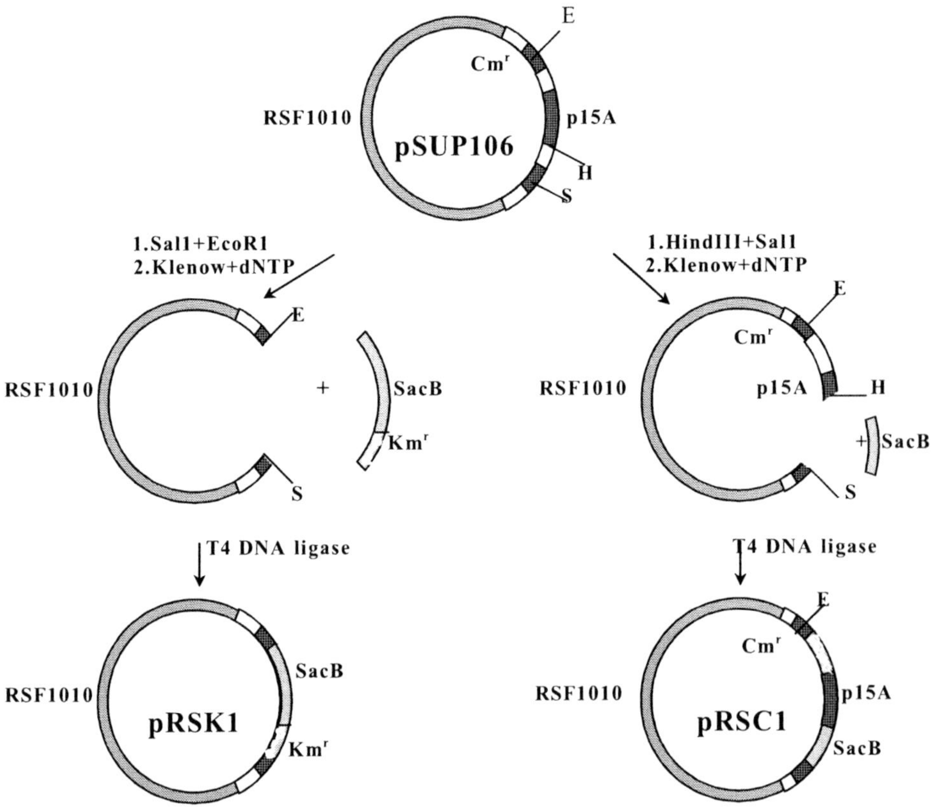

Fig. 4. Schematic of obtainment of *sac*B-bearing plasmids.

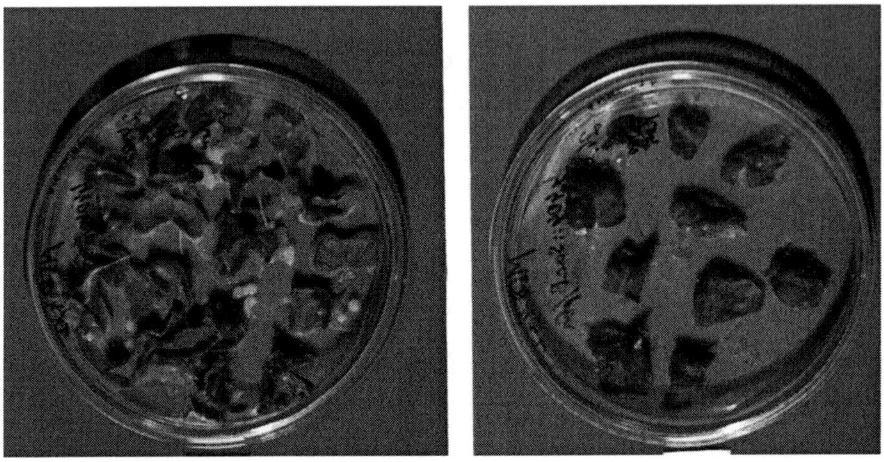

A B

Fig. 5. A. Bacterial contamination of tobacco leaf explants transformed with *A.tumefaciens* LBA4404 (pAL4404, pGA482). B. Tobacco leaf explants transformed with *A.tumefaciens* LBA4404 (pAL4404, pRSC1, pGA482) without contamination.

Plasmids were introduced by direct transformation into *A. tumefaciens* LBA4404 cells. This strain was tested in the transformation of a model plant object, tobacco (Fig. 5). The transgenic nature of the plants was confirmed by DNA hybridization methods. A change in the procedure of co-cultivation of explants with bacteria had no appreciable effect on the efficiency of tobacco transformation. A disadvantage of the method is the continuing operation of levansucrase after the death of a bacterial cell. The enzyme released from the cells keeps functioning, which may lead to the sliming of explants. Work is being carried out to produce auxotrophic mutants of the laboratory agrobacterial strains.

ACKNOWLEDGEMENTS

This work was supported by the Russian Foundation for Basic Research (grants no. 01-04-97047, 02-04-49446, 03-04-49382), the Russian Ministry of Education (grant panel 2003), and the RAS Presidium (grant from panel 6). We thank Inna Sazonova for help with pQE31-*vir*E2 preparing.

REFERENCES

[1] F. Dumas, M. Duckely, P. Pelczar, P.Van Gelder, B. Hohn: *Proc. Natl. Acad. Sci. USA.*, 98, 485-490, *(2001).*

[2] S.B. Gelvin: *J.Bacteriol.,* 181, 4300-4302, (1998).

[3] B. Tinland: *Trends Plant Sci.*, 1, 178-184, (1996).

[4] V.A. Velikov: *Biotechnologija (Rus),* 11, 14-17, (1997).

In: Biotechnology in Agriculture and the Food Industry ISBN 1-59454-119-1
Editor: G. E. Zaikov, pp. 11-15

Chapter 2

THE INFLUENCE OF PHYTOHORMONE EPIBRASSIONOLID ON THE VERTEBRAL ANIMALS IN THE EARLY ONTOGENESIS

M. A. Egorov
Astrakhan State University
The Institute of Natural science
Biological faculty
Chair of biotechnology
Russia, 414056, Astrakhan, 20-a Tatishchev st.
tel. (8512)25-17-54; fax (8512)25-17-18
e-mail logos@astranet.ru; egorov@astranet.ru

ABSTRACT

The experimental information presented in this work as a review are the results of the effects of perspective biologically active substance action-representative of brassinosteroid class-epibrassionolid of animals' organism in different level of philogenesis. Epibrassionolid, having antioxidant qualities, normalizes a series of morphophysiological models in representatives of vertebral from fish, amphibians, and mammals in the early postnatanal ontogenesis on the action of toxical agents.

Key words: biotechnology, biologically active substances, phytohormones, brassinosteroid, epibrassionolid, ontogenesis, philogenesis, unfavorable factors of environment, antioxidants.

Modern biotechnology research comprises the wide spectrum of problems, affecting the existence separate individual, as relations and dynamics of development compound over-organisms and inside-organisms structures and functions. Now this research takes a main place among other biological sciences. It's too difficult to enumerate all the main areas and aspects of research, how it affects actions of making and remaking physiological processes, functions, other reactions animals of different tacsonomical groups in the of onto-pfilogenesis (Bekhtereva, 1998). The complicated interaction of these processes with modern environment breaks constantly because of many global courses, among them the anthropological courses. The organism tries to compensate consequences of these actions using many possible means and biologically active substance of different kinds help organisms in this struggle. The studies of morphologophisiological effects and mechanisms of influence of bioactive substance on the organism of different animals is one of main problem in the modern biology (Baksht, 1991).

Anthropological interaction become the most multiactually; many numbers of sinergical effects arise but they are not much studied (Karnaukhov, 1994; Ivanov, 1997; Khodorevskaya, 1997). The powerful negative influence spreads to the hydrosphere as a soiling of air aphere, break ecological system of dry land. This factor also concerns to Volgo-Caspian basin (Ivanov at al, 1995). It needs necessity to work out, to incalcate and to use the alternative approaches in the breeding and preservation of valuable variety of animals. The usage of immunomodulates and biostimulations on the early levels of ontogenesis with purpose of growing of organisms resistance for press of unfavorable influence environment is one of the possible methods of rising preservation of valuable biological objects.

After many years of studies the generalizations of empiric data on the phenomenology of the biologically active substance Epibrassionolid (a perspective representative of the Brassinosteroids class) (Horace at al, 1990; Khripach at al, 1993; Egorov, 1995) on the morphophysiological and physiological-biochemical indexes in the Early ontogenesis (Egorov, 2002; Egorov, Vitvitskay, 2002) has been done.

It has been shown that there are no breaches in animals, organisms after experimental use f this BAS in optimal concentration (Egorov, 1998). Epibrassionolid makes for successful working-out of vertebral animals' conditioned reflexes in the Early Ontogenesis under intoxication (Egorov, Vitvitskay, 1998).

The researches were made in the laboratory of biotechnologies of Astrakhan State University (AGU), in the industrial places of Kizan sturgeon fish breeding (Astrakhan region and in the Adler industrial-experimental fish breeding salmon factory Krasnodar region).

The object in our work in different years were represented for three classes of vertebral animals: fishes (Pisces), amphibians (Amfibia), mammals (Mammalia). From many different classes of fish, this class we consided in our work on example of representatives of some varieties from three detachments – ancient, during evolution, stugeonforms (Acipenseriformes), detechment of salmonforms (Salmoniformes) and detachment of carps (Cypriniformes). The main element of this work is consideration of Epibrassionolid action on the organism from amphibians' class, where the main part was given to representative of kind Ran. The logical continuation of our work become researcher of Epibrassionolid action on the organism of lake frog (Rana ridibunda). The main question in this research is learning of biological active substance' s influence on the organisms of vertical animals. The part of experimental researches of morphophisiological analyses we were done on the animals in the laboratory.

The work concentration of Epibrassionolid we prepared according to the method of express-test (Egorov, 1998), from a part of "Epin" mixing in the 250 gr. Physiological solution, received wombihg solution with concentration 1 mg/l. Then we prepared a work solution 10^{-4} mg/l, adding it into aquariums with fishes from calculation 1 ml. water and using another methods (Egorov, 2002). The work concentration and time influence were selected for empirical way on the base of present researches (Lukyanenko, 1983, Zhuravleva, Zemkov, 2000).

According to the results of the peroxid lipids' oxidization (PLO) process research Epibrassionolid having antioxidant qualities stabilizes the penetrability of histohaematical and haemoencefalical barriers under toxical agents (heavy metals, topactive substance, oil products) influence. The BAS under study normalizes the level of ferment activity of the blood erythrocyte catalase and hemoglobin and in the some cases prevents pathological changes in parenchimatose organs of the fishes, amphibians, mammals in the early postnatal oontogenesis (Egorov, 2002; 2003). Protective qualities of Epibrassionolid (in the form of the preparation Epin) in the early ontogenesis of sturgeons, salmons, carps, under influence of unfavorable environmental factors and toxicants have real practical application in fish breeding. Now the preparation "Epin" is successfully used as a bioprotector in order to increase the level of valuable hydrobionts' survival on the Southern Russia fish-breeding plants.

The main purpose of our researches was clearing up of peculiarities of physiological effects of action biological active substance Epibrassionolid on the organisms of the representatives of different class of vertebral animals in the early ontogenesis.

In our research we were deciding the following:

1. On the base of modern methods to carry out possible main directions of influence of biological active substance – brassinosteroid on the organism of different class animals.
2. To characterize peculiarities of behavior reactions for Epibrassionolid for young white sturgeon, Russian sturgeon, salmon of Black Sea, carp and another fishes also laboratory rates in the norms and in the unfavorable conditions.
3. To carry-out physiology and biochemistries peculiarities of Epibrassionolid influence of fishes, amphibians, mammals in following indexes: level of oxide peroxide of lipids, possible of penetrate gematoencefalical and some of gystomatical barriers and hemoglobin level of blood.
4. To learn and character the main multifunctional changes in fishes' liver, amphibians mammals in the early ontogenesis in the action of phitogormon of Epibrassionolid, exzotoxicates and also include biological active substance with the influence of unfavorable factors on the animals organisms.

In our research we used a complex way for learning action of Epibrassionolid on physiological condition of animals in the early ontogenesis. They are physiological, morphological and biochemistry (Hetru et al., 1986; Richter et al., 1987). All of results we cultivated from statistic method (Dahse et al., 1990; Ugarova, 1994). On the base of this researches we known some of pecularities of action biological active substance. Epibrassionolid has some resemblance properties as a structure of steroid gormon of animal and plant origin. This fact indelined the cammon nature of steroid gormous of this row animal and plant origin (Ugarova, 1994; Egorov, 1998). The brassinosteroid penetrate into the organisms of unfavorable (Khripach et al., 1997), causing different specifically and general effects (Egorov, Vitvittskaya, 2002; Kolman et al., 2001). The industry of epibrassionolid is developing and year from year it become more modern and substances – more active (Kojiro et al., 1984, 1985). The different part of substance, on the base of which is epibrassionolid need in express-test on the biology objects. The selection of object can be connection with real changes in the level morphophisiological index and possibilities of conducting of statistical analyses (Egorov, 2002).

We propose some different express-test of evaluation of active new substance. It was so possible for understanding of this biology active substance on the influence unfavorable factors of environment of animals. For example, express-test for fishes: sturgeons, salmons, carps. We conducted some of experiments for evaluation of different substance on the toxical action of dietanolamin. We were shown that the method of test biological active of different substance was really base (Egorov, 2002; Vitvittskaya, Egorov, 2002).

The results of this work are about possibilities of practical usage of biological active substance-epibrassionolid as substance "Epin" in a fish breeding factory. Now this substance is used in the Astrakhan region and Krasnodar region with the purpose of rising fish production.

REFERENCES

[1] A.V. Karnaukhov: For question about chemical structure of atmosphere and thermal ballans of Land // *Biophisica*, T. 39, N1, 1994. P. 148-152.

[2] C. Hetru, J.P. Roussel K. Mori, Y. Nakatani: Brassinosteroids // *C.R. Acad. Sci. Ser.* 2. 1986. Vol. 302, №7. P. 417-420.

[3] G.C. Horace, Y. Takao, A. Gunter: Brassinosteroids: Chemestry, Bioactivity, and Applications // *ACS symposium series*, 474, Washington, DC, 1991.

[4] G.F. Zhuravleva, G.V. Zemkov: The structure breaches in the ootsitos in depend on patomorphogenesis of sturgeon liver for cumulative toxikoses // *International science conference. The thesises. "Sturgeones in the XXI century"*. Astrakhan, 2000. P. 144-145.

[5] Dahse., C.M. Willmer G. Graamatikopoulos, D. Nikolopoulos: New effects of brassionolid // *Int. Conf. on Brassionolids: Abstracts*. 1990. P. 34.

[6] K. Richter, G. Adam, H.M. Vorbort: Class of Brassinosteroids // *J. Appl. Entomol.* 1987. Vol. 103, №5. P. 532-534.

[7] L.V. Vitvitskaya, A.M. Tihomirov: Influence of different biologically active substances on stellate sturgeon survival in ontogenesis stages // *3-rd International symposium on sturgeon. Booklet of abstracts.* Piacenza, Italy, July 8-11. 1997. A2-P.

[8] L.V. Vitvitskaya, M.A. Egorov: The antiterratogenical effect of biological active substance and perspectives their usage in the sturgeonbreeding // *The thesis XXXX science conference* AGTU, Astrakhan, 1996. P.81-82

[9] m.a. egorov, l.v. vitvitskaya: The evaluation of biologically active substances influence on sturgeon embriogenesis under unfavorable conditions // *The Caspian sea science education technologies. International journal of collected academic articles*. Book I. Astrakhan. 1998. P. 104-109.

[10] M.A. Egorov, L.V. Vitvitskaya: *The usage of biology active substance in the artificial breeding of sturgeon Volgo-Caspean region*. Astrakhan: AGPU, 2002. P. 100.

[11] M.A. Egorov, U.V. Altufiev, N.S. Altufieva: The development of morphofunctional consideration of liver lake frog's change (Rana ridibunda) // *All Russian scientific*

conference "Ecological and biological problems of Volga region and Northern Caspian. Astrakhan: AGPU, 2002. P. 190-191.
[12] M.A. Egorov, V.P. Zagriichuk: New directions of researches and some of practical ways of usage of biological active substance of Epibrassionolid in the fishbreeding //J. *"Science to industry",* N6. Moscow, 2001. P. 15-17.
[13] M.A. Egorov: *Physiological peculiarities of epibrassionolid phytogormon action on the organism of animals in the early ontogenesis.* Astrakhan: AGPU, 2002. P. 269.
[14] M.A. Egorov: The antiterratogenical action of some biological active substance on young sturgeon in the unfavorable ecological condition // *International science conference Ecology and region.* Rostov, 1995. P. 93.
[15] M.A. Egorov: The sperms sturgeon as a test-object for express evaluation of biostimulates activity // *International science conference. The thesises.* Astrakhan, 1998. P. 135-136.
[16] R. Kolman, M. Shchepovsky, G. Kolman: The influence of epibrassionolid on the growing of young besters // *The thesises. International conference. "Aquaculture sturgeon.* Astrakhan, 2001. P. 100-101.
[17] R.P. Khodorevskaya: The valuation of sturgeon stocks in the Caspian sea // *Fishbreeding.* N5. 1997. P. 39.
[18] T.U. Ugarova: *The family vegetablegrowing in the narrow rows.* Samara, 1994. 160 p.
[19] T.V. Bechtereva: *Ecological and physiological researches.* P. 121.
[20] V.A. Khripach, F.A. Lakhvich, V.N. Zhabinsky: Brassinosteroids. Minsk, "*Science and technique*", 1993. P. 287.
[21] V.A. Khrpach, V.N. Zhabinski, N.N. Malevannaya: Recent advances in Brassinosteroids study and application // *The plant growth regulation society of America.* Atlanta, Georgia August 8-12, 1997, P. 101-16.
[22] V.I. Baksht: Noosphere // *Noosphere.* №1. – 1991.
[23] V.I. Lukiuanenko: *The general ikhtiotoxiology.* M., 1983. P. 320.
[24] V.P. Ivanov, A.D. Vlasenko, R.P. Hodorevskaya: The ways of sturgeon preservation // *Fishbreeding.* N2. 1995. P. 24-26.
[25] V.P. Ivanov: The condition and perspectives of row material and extraction of fish in Caspian sea // *Congress of Russian ikhtiologers of Russia. The thesises.* M., House-keeping VNIRO, 1997. P. 421.
[26] W. Kojiro, K. Hiroaki, M.A. Shingo: Simple Bioassay for Brassinosteroids: A Wheat Leaf-unrolling test //*Agric. Biol. Chem.*, 49 (7). 1985. P. 22-49.
[27] W. Kojiro, M.A. Shingo, A. Hiroshi: A Rice Lamina Inclination Test – Micro-quantitative Bioassay for Brassinosteroids // *Agric. Biol. Chem.,* 48 (3). 1984. P. 719-726.

In: Biotechnology in Agriculture and the Food Industry
Editor: G. E. Zaikov, pp. 17-22
ISBN 1-59454-119-1

Chapter 3

OPTIMIZATION OF PROSO MILLET (*PANICUM MILIACEUM* L.) CLONING

S.V. Bobkov
The All-Russia Research Institute of Legumes and Groat Crops,
p/b Streletskoye, Orel, 302502, Russia
e-mail: office@vniizbk.orel.ru
fax: (0862)403130

ABSTRACT

The authors investigated the determinants of regenerating callus tissues of millet during long-term cultivation. A way of optimization callus tissues transfer on fresh media with adaptation of regenerated plantlets with roots into non-sterile conditions was developed. The formalization of dynamic characteristics in long-term cultivated *in vitro* regenerating callus tissues was shown, that permit developing on this basis quantitative models for cloning of millet plants.

Key words: proso millet, anther culture, callus tissues, long-term regeneration, proliferation

Embryogenic callus tissues and regenerated plants of millet (*Panicum miliaceum* L.) have been obtained in cultures of immature inflorescence and embryos, mature seeds, segments of leaves and stalks at use as growth regulator of 2,4-dichlorophenoxyacetic acid (2,4-D) (1, 2). Regeneration process was observed after transference of embryogenic callus tissues on mediums without growth regulators or with low content of 2,4-D, but it proceeded during a short period of time. Successive transfer of embryogenic calli to mediums R1 and MC ensured continuous regeneration of plantlets and proliferation of callus tissues (3, 4). At optimal combination of these processes which took place in medium conditions of MC, duration of callus tissue cultivation *in vitro* was essentially enlarged, that allowed transfer of regenerated plantlets into non-sterile conditions during long period of time, calculated by years. In this connection there was a necessity for analysis of regenerating callus tissues parameters in dynamics, that which the purpose of the present work.

Methods

Object of research were three long-term cultivated callus clone, obtained in anthers culture of millet hybrids F_3: gm1 (Mohar-like panicle 2009 × Sorghum-like dwarf 2010); aml29 (Regenerated plant 20№10 × Sorghum-like dwarf 1959); am26 (Sorghum-like dwarf 2010 × Mohar-like panicle 2009). Callus clones have been obtained on the mediums, containing salt mediums MS (5), vitamins of medium B5 (6), L-glutamine or L-glytamic acid (500 mg/l), glycine (2 mg/l), myo-inositol (500 mg/l) and 2,4-D (2 mg/l). For induction of regeneration process there was used medium R1, containing myo-inositol, glycine, 6-benzylaminopurine (6-BAP), α-naphthalene acetic acid (α-NAA) in concentration correspondingly 100, 2, 10, 0,5 mg/l, and also salts of medium MS and vitamins of medium B5. Regenerating callus tissues were cultivated on medium MC (salts MS, vitamins B5, without growth regulators) with periodic transfer of regenerated plantlets into non-sterile conditions (3). Duration of cultivation on medium MC for clones gm1, am129 and am26 made accordingly 895, 775 and 805 days. On 5, 10, 15, 20, 25, 30, 35, 40, 50th days of cultivations *in vitro* determined mass and volume of callus tissues, and also number of plantletss- which length was more than 1,5 cm. For statistical analysis of the data used the computer programs of correlation analysis and ANOVA: two factors without replications.

Results

Between mean values of mass and volume of callus tissues, and also number of plantlets of gm1, am129, am26 clones- in dependence on time of cultivation on medium MC were the high degree of correlation (table 1. On the 20th day after replanting of regenerating callus tissues segments on fresh mediums, parameters of all three attributes have essentially increased; maximal values were marked on the 35th day of cultivation, then the mass and volume of callus tissues were in limits of statistical error. The number of plantlets was decreased.

Table 1. Mass, volume of callus tissue and number of plantlets of millet in dependence on time of cultivation on medium MC (mean values of gm1, am129, am26 clones)

Time of cultivation in vitro, days	Mass of callus tissue, g	Volume of callus tissue, cm^3.	Number of plantlets
5	0.07	0.13	0.18
10	0.16	0.45	0.30
15	0.28	0.54	0.42
20	0.56	1.33	0.49
25	0.93	2.43	1.00
30	0.91	2.71	0.58
35	1.93	5.10	2 17
40	1.73	2.71	1.48
50	1.69	4.45	1.51
LSD_{05}	0.32	0.72	0.39

Note: $r_{mass\text{-}volume} = 0{,}96$; $r_{\text{mass-number of plantlets}} = 0{,}93$; $r_{\text{volume-number of plantlets}} = 0{,}94$.

On the 35th day of cultivations the average mass and volume of callus tissues made accordingly 1,93 g and 5,1 cm^3. For every glass on the average it was 2,17 plantlets- with length more than 1,5 cm, suitable for transfer in non-sterile conditions. Differences between regenerating callus clones were revealed in number of plants and volume of callus tissue (table 2). By mass of callus tissues clones did not differ.

Table 2. Mass, volume of callus tissue and number of regenerated plants of millet in dependence on clone

Clone	Mass of callus tissue, g	Volume of callus tissue, cm^3.	number of plantlets
gm1	0.92	2.3	0.69
am129	0.88	2.8	0.74
am26	0.91	1.5	1.28
LSD_{05}	0.19	0.42	0.23

On the basis of dynamic mean values of three millet clones (gm1, am129, am26) we had developed the optimal schema of cultivation and replanting into non-sterile conditions with the purpose of deriving the greatest number of plantlets with the least expenditures of time and expendables (table 3).

Table 3. The schema of replanting of regenerating callus tissues of millet plants into non-sterile conditions

Time of cultivation in vitro, days	Number of plantlets	Number of 5-day periods	Average number of plantlets
5	0.18	1	0.18
10	0.30	2	0.15
15	0.42	3	0.14
20	0.49	4	0.12
25	1.00	5	0.20
30	0.58	6	0.10
35	2.17	7	0.31
40	1.48	8	0.19
50	1.51	10	0.15

Research of dynamics of change of number of plants was performed with a 5-day interval, and into account was accepted number of 5 day terms, falling to the time of cultivation of callus tissues *in vitro* (5, 10, 15, 20, 25, 30, 35, 40, 50 days). Counted number of the plants obtained during 5 days of cultivation in each variant: the maximal number of plantlets reflected optimal time for replanting (see table 3). The greatest number of plantlets, obtained in recalculation for 5-day term, fell on the 35th day of cultivation. Thus optimal time of cultivation before replanting of callus tissues makes 35 days. If from one glass for replanting take two segments of regenerating callus tissue, it is possible to determine number of (u_n) on the *n-th* stage of cloning. Under these conditions, dependence between parameters it is possible to introduce as geometrical progression: u_n = 2,17 2^{n-1}. If we take into account all the regenerants, present for the beginning of cloning, then the generalized formula becomes $u_n = aq^{n-1}$, where a — number of regenerants, suitable for transfer into non-sterile conditions at the beginning of cloning, q — number of segments from one regenerating callus tissue. According to the rule of summarize of members of geometrical progression we find total number of regenerants obtained on *n-th* stage of cloning:

$$S_n = u_1 + u_2 + u_3 ... + u_n + ..., \text{ or}$$

$$S_n = \frac{a(1-q^n)}{1-q}. \qquad [1]$$

Transform this formula with the purpose of definition of necessary amount of stages of cloning for deriving of the planned number of regenerants:

$$S_n(1-q) = a(1-q^n);$$

$$\frac{S_n(1-q)}{a} = 1 - q^n;$$

$$q^n = 1 - \frac{S_n(1-q)}{a} = \frac{a - S_n(1-q)}{a}; \qquad [2]$$

$$n = \frac{\ln[a - S_n(1-q)] - \ln a}{\ln q}.$$

Stages of cloning are represented with all members of geometrical progression, including initial clone material. Hence, the number of replantings *(p)* is one unit less than number of members:

$$p = n - 1, \text{ or}$$

$$p = \frac{\ln[a - S_n(1-q)] - \ln a}{\ln q} - 1, \qquad [3]$$

As a rule, the number of replantings designed under the formula [3], is not the whole, therefore for deriving the set number of regenerated plants fulfill all complete replantings, and for last take only a part of available material.

Initial number of regenerants (a_1) for performing of the last incomplete replanting is determined by the following formula:

$$a_1 = \frac{S_n - S_{nn}}{q}, \qquad [4]$$

where S_n — planned number of regenerated plants at cloning stage $n = p + 1$, S_{nn} — sum of regenerated plants on the last complete cloning stage $n = p + 1$, where p — whole number. S_{nn} is determined by formula [1]. For last replanting take such amount of units of cultural glasses in which it would be introduced number of regenerated plants equal to a_I.

EXAMPLE

If we assume, that at the beginning of cloning 5 glasses are available, then initial number of regenerants a=2,17x5=10,85. If number of segments of one regenerating tissue, transferred to fresh mediums q=2, it is possible to calculate amount of regenerant plants (u_n) at different stages (n) of cloning, their sums (S_n) and to compound the obtained data with number of replantings (p) and length (t) of cloning (table 4). If it is necessary to obtain 1000 regenerant plants with condition, that a= 10,85 and q=2, then number of replantings, calculated by formula 4, will equal to:

$$p = \frac{\ln[10{,}85 - 1000(1-2)]\ln 10{,}85}{\ln 2} - 1 = \frac{\ln 1010{,}85 - \ln 10{,}85}{\ln 2} - 1 = 5{,}5417$$

It means, that 5 complete replantings will be carried out. Then calculate integral number of regenerant plants S_{nn} at n=p+1=5+1=6 cloning stage by formula 2:

$$S_{6n} = \frac{10{,}85(1-164)}{1-2} = 683{,}55$$

Using the formula 5, obtain initial number of regenerants for the last incomplete replanting:

$$a_1 = \frac{1000 - 683{,}55}{2} = 158{,}225$$

Taking into account, that one glass, on the average, contains 2,17 regenerant plants, calculate number of used glasses:

158,225:2,17≈73

So, process of obtaining of regenerated plants in anthers culture millet includes three stages: shaping of embryogenic callus tissues; stimulation of intensive regeneration process; long-term cultivation of regenerating callus tissues on the medium without growth regulators (3). Thus for clones gml, am129 and am26 average duration of first two stages makes 75 days, and the term of long cultivation of continuously regenerating callus tissues - 825 days. Duration of the third stage under with reference to all the term of cultivation occupies 91,7 % of time. Practically all regenerated plants are selected at a stage of long-term cultivation of regenerating callus tissues. Thus transfer of rooted regenerated plants into non-sterile conditions coincides with transfer of segments of regenerating callus tissues into fresh mediums.

Table 4. Efficiency of millet cloning (a=10,85, q=2, interval of replantings - 35 days)

Cloning stage, n	Number of replantings, p	Time of cloning, t (days)	Number of regenerated plants on n-stage of cloning, u_n	Summary number of regenerated plants, S_n
1	0	0	10.85	10.85
2	1	35	21.7	32.55
3	2	70	43.4	75.95
4	3	105	86.8	162.75
5	4	140	173.6	336.35
6	5	175	347.2	683.55
7	6	210	694.4	1377.95
8	7	245	1388.8	2766.75
9	8	280	2777.6	5544.35
10	9	315	5554.2	11098.55

The offered way of optimization of replantings of regenerated plants into non-sterile conditions is based on recomputation of number of regenerants on a fixed time term. Recomputation can be conducted also for one day of cultivation, that basically does not change essence of an affair. The maximal number of regenerated plants, falling for a time unit, reflects optimal time of replanting. The developed formula for definition of number of replantings, proceeding from the planned number of regenerated plants allows to envelop all the term of cultivation of regenerating callus tissues. Necessary conditions at use of this formula is presence of the data about an initial amount of regenerants and number of the replaced segments per one regenerating callus tissue.

Thus, we showed possibility of formalization of dynamic parameters of long-term cultivated *in vitro* regenerating callus tissues that allows to develop quantitative models of millet plants cloning on this basis.

REFERENCES

[1] J.W., Heyser M.W. Nabors. Regeneration of Proso Millet from Embryogenic Calli Derived from Various Plant Parts. *Crop science,* 22 (5), 1070-1074 (1982).

[2] T.S Rangan., I.K. Vasil Somatic Embryogenesis and Plant Regeneration in Tissue Cultures of *Panicum miliaceum* L. and *Panicum miliare* Lamk. *Z. Pflanzenphysiology.*, 109: 49-53, 1983.

[3] S.V. Bobkov. Deriving of rooted regenerants in anthers culture of millet. / *Vest. RASHN.* 5, 41-43 (2000). (In Russian)

[4] S.V. Bobkov. Deriving of rooted regenerant plants of millet in culture of immature racemes and anthers. *Agric. boil.*, 5, 65-68 (2002). (In Russian)

[5] T. Murashige, F. Skoog A Revised Medium for Rapid Growth and Bioasseys with Tobacco Tissue Cultures. *Physiology of Plant.* 15(13), 473-497 (1962).

[6] O. Gamborg, D.E. Eveleigh: Culture Methods and Detection of Gluco-nases in Cultures of Wheat and Barley. *Can. J. Biochem.* 46(5) 417-421 (1968).

In: Biotechnology in Agriculture and the Food Industry ISBN 1-59454-119-1
Editor: G. E. Zaikov, pp. 23-33

Chapter 4

NEW PROBIOTIC "BIOKONS" FOR PREVENTIVE SALMONELLA AND E. COLI ASSOCIATED INFECTIONS AT A POULTRY FARM

G. Dudikova, L. Tsurkan, A. Chizhaeva, T. Orljuk, Z. Tulemisova, A. Sultanov and K. Tulemisova
The National Academic Center of Agrarian Research RK
480091, Almaty, Ablay khan av., 79
E-mail:galina@niipp.samal.kz
Phone/fax: +73272573922

ABSTRACT

A new probiotic is created on the basis of the consortium strains of lactic acid bacteria of genus *Lactobacillus*: ***L. brevis - 67, L. casei var alactosus-22, L. fermentum-104 and L. plantarum-2,*** being an active inhibitor of Escherichia coli and Salmonella typhimurium. Antagonistic properties of the consortium are caused by synthesis biologically active peptides, which show high activity in the sour environment. Introduction of the probiotic to chickens daily within the first 10 days of life in the certain dose on a background of infection with salmonelloses and colibacterioses results in a 100 % medical effect. The lactic acid bacteria included in the preparation are not pathogenic, possess adhesive properties, positively influence the normal flora gastroenteric path of chickens and promote mastering of nutrients, raising the daily average gain of the live weight of poultry.

Key words: Probiotic, *Lactobacillus*, consortium, antagonism, peptide, salmonelloses, colibacterioses, pathogens

INTRODUCTION

Serious economic damage to poultry farming all world carry salmonelloses and colibacterioses of birds. Known preparations for protection of young growth of agricultural birds against infectious diseases, such as vaccines, bacteriophages, antibiotics are not always effective [1]. Therefore, now, for restoration broken microbiocenoses birds use various receptions, first of all introduction of probiotics.

Probiotics, used for animals include *Lactobacillus, Bifidobacterium, Bacillus, Streptococcus, Pediococcus, Enterococcus,* and yeast of genus *Saccharomyces* [2-7]. Use *Lactobacillus* as probiotic for animals is expedient as these bacteria are representatives of normal microflora of intestines of animals, birds and the person; render positive effects on physiological, biochemical and immune reactions of an organism of the owner [8].

It is known, for example, that probiotic preparations for the prevention colibacterioses at the chickens, prepared of a monoculture of the intestinal sticks allocated from hens, or *Lactobacillus* and *Propionibacterium*, allocated from a sheep. For the prevention of a salmonellosis and campylobacterioses of young growth of agricultural birds and animals it is offered to use for feeding monocultures of *Lactobacillus* or *Enterococcus* [9,10].

It is necessary to note, that, despite of the certain successes of research in the area of probiotics, spent in Kazakhstan and in other countries of near and far abroad, it has not produced uniform approaches to creation steady in structure of probiotic the forms of microorganisms, and also development of "know-how" and applications of such preparations. In the given article results of research on creation new probiotic are presented on the basis of the consortium *Lactobacillus* for treatment salmonelloses and colibacterioses at birds.

MATERIALS AND METHODS

Bacterial Strains and Culture Media

The inhibitor strains were *Lactobacillus plantarum 2(B90), L. casei var. alactosus 22(B62), L. brevis 67(B54), L fermentum 104(B64)*, consortium of four listed above strains, and their mixture prepared from overnight culture. Lactobacillus strains were isolated from wheat flour and obtained from the Kazakh Scientific Research Institute of Food Industry.

The model bacterium used in the determination of inhibition was *Escherichia coli TG-1/K12* (Cambridge Antibody Technology, Melbourn, United Kingdom), *Salmonella typhimurium* (Oak Ridge National Laboratory, Oak Ridge, TN, USA).

Bacterial Growth

Lactobacillus strains were grown in de Man-Rogosa-Sharpe (MRS), broth (BBL) at 37°C for 24 h. The E. coli strain was cultured in Luria-Bertrani (LB) medium (Difco) at 37 °C overnight. The Salmonella typhimurium strains was cultured in beef-extract broth, Brilliant-green Phenol-red Lactose Sucrose Agar (BPLS Agar) at 37 °C overnight.

Analysis of Process Periodic Cultivation

Separate strains of Lactobacillus: L.plantarum-2, L. fermentum-104, L. brevis-67, L. casei var. alactosus-22, included in the Consortium, it has been carried out scale cultivation on medium of the following composition: glucose – 20 г; peptone – 10 г; KH_2PO_4– 2,0 г; NaCl– 5,0 г; yeast autolysate – 50,0 г; acetate of sodium – 2,0 г; sulfate of magnesium – 0,1 г; the Twin - 80 – 1,0 г; water – the rest up to 1 litre. An one-daily inoculum brought in quantity of 10 %. Supervision carried out during 48 h, sampling for the analysis selected with an interval - two hours.

Dynamics of Relation of the Cultures Making a Consortium

Definition of dynamics of relation of the cultures making a consortium, we had been applied a method of luminescent antibodies (indirect way Kuns). Each of cultures was painted by immune homological serums, then luminescent whey. Calculation of cells was carried out under a luminescent microscope "Axioskop 40" FL.

Assays for the Inhibitory Activity of Lactobacillus Strains

Screening of bacteriocin production was performed using a multilayer agar plate system which consisted of the following four separate agar layers in 12 cm petri dish (i) a supporting layer of 10 ml MRS agar (1.4% agar in MRS broth) at the bottom, (ii) a layer of 10 ml MRS soft agar (0.7% agar in MRS broth) containing the bacteria to be screened, (iii) a layer of 10 ml LB soft agar. The plates were incubated at 37 °C for 48 h to allow time for bacteriocin production. Finally the plates were overlaid with a fourth layer of 10 ml LB soft agar containing the indicator organism (OD600 0.35) and incubation is continued for 18 h to develop growth inhibition zones. This technique was modified from D.B.Diep et al [11].

Comparison the inhibitory activity of different strains was performed using the modified method of plate assay. E. coli strain was grown overnight and overlaid onto plates previously prepared with a bottom layer (20 ml) of LB medium containing 1.4% agar. Holes (9 mm) were punched aseptically in the agar using a cork borer. Lactobacillus culture, grown overnight in the MRS broth, was placed (100 (l) in the appropriated wells. Test samples were kept in the refrigerator for 2 h to allow the diffusion of inhibitory compounds into agar. After that the plates were incubated for 18 h at 37ºC [12].

Determination of Minimum Inhibitory Concentration

The cell-free culture supernatants from different Lactobacillus strains were prepared from overnight cultures by centrifugation at 4000 **g** for 15 min at 4 °C and sterile filtration through a 0.45 μm HA filters (Millipore Corp., Bedford, MA, USA). Supernatant was diluted in LB medium for the determination of minimum inhibitory concentration. All samples contained diluted supernatant and the indicator strain (1% v/v) were incubated for about 18 h at 37 °C, after which growth inhibition of the indicator organism was measured spectrophotometrically at 600 nm.

Studying Adhesive Properties LAB

Research of adhesive properties of lactic acid bacteria carried out in vitro, using a surface epithelium cells of intestines of white mice by the standard technique [13].

Initiation of Probiotic Performance Efficiencies

Definition of efficiency of action of probiotic have been continued concerning pathogenic and conditional - pathogenic microflora. In quality of probiotic used the lyophilic dried preparation on the basis of consortium of Lactobacillus. For infection of chickens used virulent culture Salmonella typhimurium (USA).

From the date of receipt from an incubator, the chickens water of "Probiotic" in volume 1ml through a mouth during from one till 5 days, the titer made 10^6.

Then chickens infected with culture Salmonella typhimurium in a doze of 10 m. b./1 ml in volume of 1 ml through a mouth. Within 5 days after infection added in potable water of probiotic which doze contained 10^7 cells LAB.

Then chickens were is killed, opened and made crop of contents of intestines on beef-extract broth and Brilliant-green Phenol-red Lactose Sucrose Agar (BPLS Agar).

In this article the following variants of experience were used:

- the control 1(positive) – intact chickens;
- the control 2 (negative) - the chickens infected Salmonella, not receiving treatments;
- 1 group – the chickens infected Salmonella, receiving treatment with probiotic within 5 days;
- 2 group – the chickens infected Salmonella, receiving treatment with probiotic within 2 days

RESULTS

Relation of Lactobacillus Cells in Consortium in Process of Cultivation

Supervision over dynamics of change of the specific relation of cultures in a consortium have shown, that the relation of cultures of a consortium varies during all 48 hours of cultivation (Figure 1). So, at inoculation (0 hours), the relation of cultures makes 1:1:1:1. After 12 hours of joint growth, culture *Lb. plantarum 2* starts to prevail of other cultures and by 24 hours of cultivation its quantity makes all 2/3 cells of a consortium. After 36 hours of cultivation the share of cells *Lb. plantarum 2* in a consortium starts to decrease and by 48 hours the relation of cultures again comes back to value - 1:1:1:1.

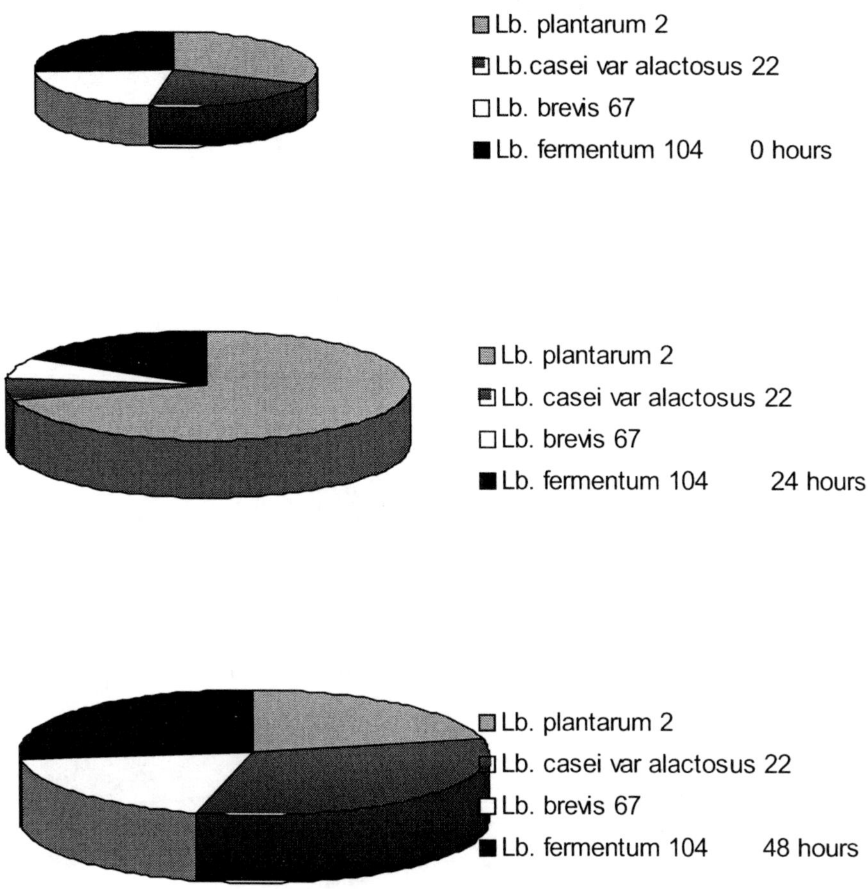

Fig.1 Dynamics relation of Lactobacillus cells in consortium in process of cultivation

Bacteriocin Production

Under standard cultivation conditions four *Lactobacillus* strains produce antibacterial compounds constitutively during exponential and early stationary-growth phase. Furthermore, when fresh inoculum from an overnight culture was diluted to a final cell density of approximately 10 CFU ml^{-1} the inhibitory activity was also found. It was shown that each colony in diluted *Lactobacillus* culture could produce proteinaceous compound(s) that inhibits the *E.coli* TG1 growth (Fig. 2). The meaning that these compounds are proteins was confirmed by adding the papain (0.264 mg ml^{-1}) in supernatant of each culture, mixed culture and consortium supernatant. The presence of this enzyme in *Lactobacillus* supernatant lead to losing of its inhibitory activity. However, papain (0.264 mg ml^{-1}) in the third layer in multi-layer assay (see methods above) did not affect the inhibitory activity (Fig. 2 B). These data suggest that papain, does not affect *Lactobacillus* cells and their ability to produce bacteriocins.

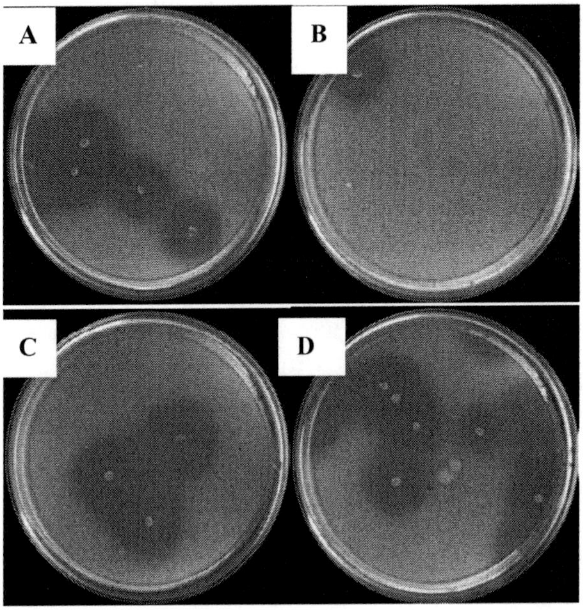

Fig.2 Bacteriocin production by four different *Lactobacillus* strains Plate A - *L. plantarum* 2(B90); plate B - *L. casei var. alactosus* 22(B62); plate C - *L. brevis* 67(B54); plate D - *L fermentum* 104(B64). Plate B - the third layer contains papain (0.264 mg ml^{-1}) in LB soft agar. Plates A, C, and D contain no enzyme in the third layer.

Determination of Minimum Inhibitory Concentration

The cell-free culture supernatant from four separately grown *Lactobacillus* strains and consortium was diluted with LB medium in ratios of 1:1 to 1:45. The minimum inhibitory concentration was defined as the amount of *Lactobacillus* supernatant in LB medium that inhibited growth of the indicator organism by 50% (50% of the turbidity of the control culture without *Lactobacillus* supernatant). It was found that supernatant after co-cultivation of *Lactobacillus* strains exerted a larger effect than individual strains alone. Firstly, it seemed that higher inhibitory activity occurred due to a synergy effect between all compounds that the strains produced. A similar effect was expected when supernatant from overnight cultures was mixed in equal amounts. Fig. 3 shows the difference between inhibitory activity of these samples. Inhibitory activity increased two-fold in consortium in comparison to activity of mixed culture (20-fold dilution and 10-fold dilution respectively decreases *E. coli* TG1 growth by 50%).

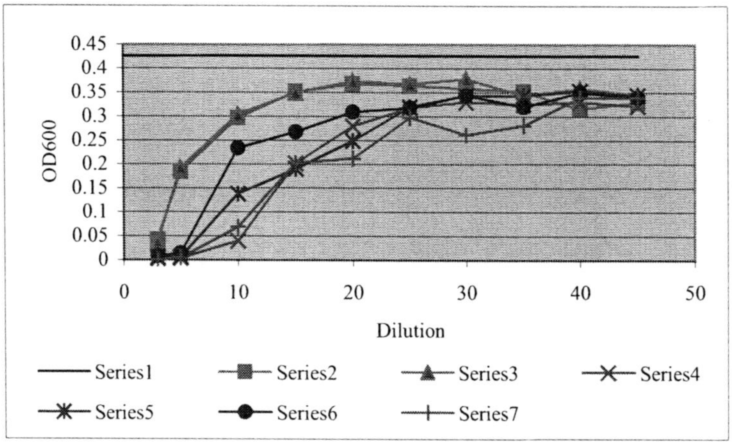

Fig. 3 Influence the concentration of *Lactobacillus* supernatant in LB medium on *E. coli* TG 1 growth. Series 1 – OD_{600} of *E.coli* TG1 in pure LB medium; series 2 – *L. plantarum* 2(B90); series 3 – *L. casei var. alactosus* 22(B62); series 4 – *L. brevis* 67(B54); series 5 – *L. fermentum* 104(B64); series 6 – mixed culture; series 7 – *Lactobacillus* consortium. Dotted lines indicate decreasing of *E.coli* TG1 growth by 50%.

Storage of Lactobacillus Supernatant

As shown, co-cultivation of four *Lactobacillus* strains lead to greater inhibitory activity. This effect was not lost during storage for 30 days at 4 °C (data not shown). In case of storage at room temperature and 37 °C, the inhibitory activity of consortium's supernatant was stable for 7 days only (Fig. 4B, 4C). In addition, the inhibitory activity of the consortium's supernatant was higher when all samples were kept for 7 days at higher temperatures: 50 °C (Fig. 4D) and 75 °C (Fig. 4E). All samples – supernatants of individual *Lactobacillus* strains and consortium – inhibited *E. coli* TG1 growth even after autoclaving during 30 min at 121 °C. An average decreasing of inhibitory activity after autoclaving was 32.83%.

Adhesive Properties

The tested lactic acid bacteria *in vitro* are fixed on the surface intestinal epithelium cells of white mice with a various degree of adhesiveness. Given table 1 show, that the degree of sticking at researched microorganisms depends on species of lactic acid bacteria, higher index of adhesiveness (3,34) possesses *Lactobacillus plantarum 2*, and low (2,96) - *Lactobacillus brevis-67*.

Initiation of Probiotic Performance Efficiencies

Apparently in figure 5, the chickens receiving of probiotic within 5 days after infection Salmonella typhimurium (1 group) typed in weight better. The difference in weight with chickens of other groups made 5-8 gramme. Chickens of 1 group were larger, more active, had good appetite, equal and dense down. Chickens of control groups were less size, lagged behind in weight, were languid, passive, had rare plumage.

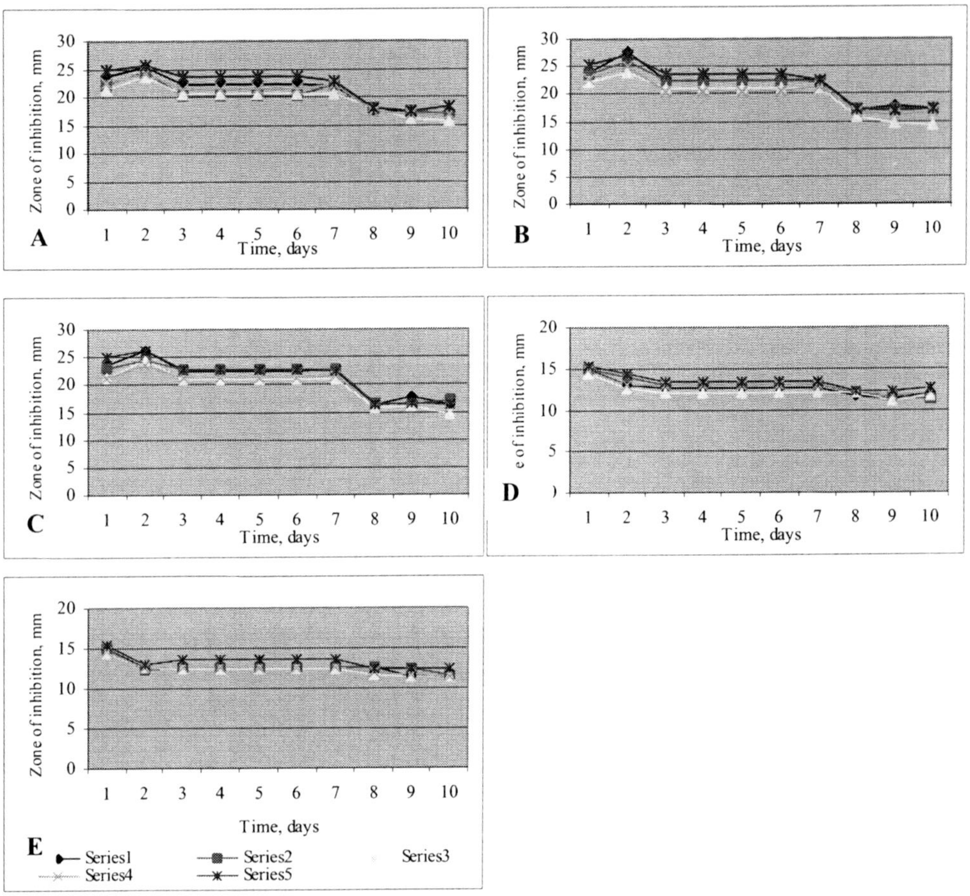

Fig. 4 Storage *Lactobacillus* supernatant in five different temperature A – 4 °C; B – 20 °C; C – 37 °C; D – 50 °C; E – 75 °C Series 1 – *L. plantarum* 2(B90); series 2 – *L. casei var. alactosus* 22(B62); series 3 – *L. brevis* 67(B54); series 4 – *L. fermentum* 104(B64); series 5 – *Lactobacillus* consortium.

Table 1 Adhesion of lactic acid bacteria to intestinal epithelium

№	Strains of LAB	Index of adhesiveness of bacteria
1	*Lactobacillus plantarum-2*	3,34
2	*Lactobacillus fermentum-104*	3,14
3	*Lactobacillus brevis-67*	2,96
4	*Lactobacillus casei var. alactosus-22*	3,22

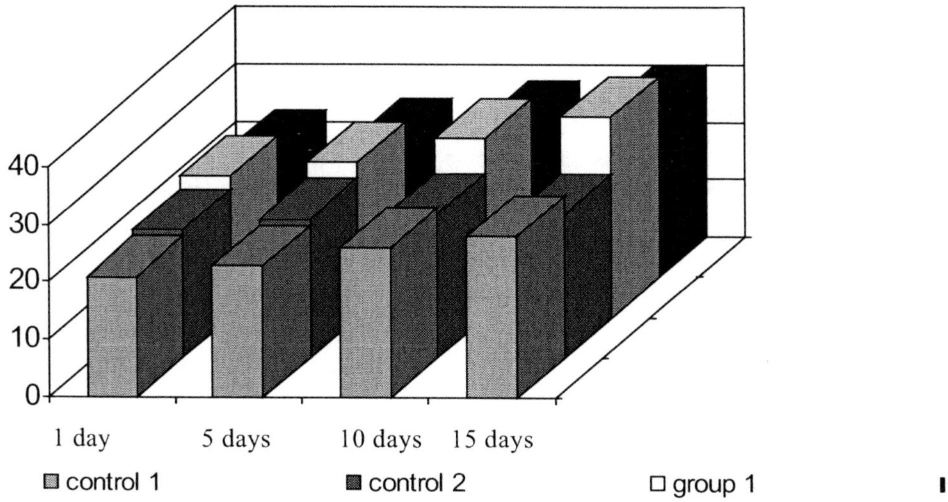

Fig. 5 Dynamics of an additional weight at the chickens receiving treatment of salmonelloses with the help of preparation "Biokons" Axis of abscises - duration of cultivation of chickens, days; axis of ordinates – weight of a body of chickens, g

The chickens of 2 groups receiving of probiotic within 2 days, nevertheless, well typed weight and were more active than chickens of control groups.

After killing chickens in 1 and 2 groups it was not observed a characteristic pathalogo-anatomic picture of disease by a salmonellosis. Some chickens had a hyperemic liver, however the internal bodies were without changes, intestines in norm and contain plenty of cells of lactic acid bacteria (titer 10^3). After five days of treatment with a probiotic, in intestines of chickens of the first group did not reveal cells of salmonellas. In the intestines of the second group of chickens individual cells of salmonellas have been isolated.

DISCUSSION

In the present study we have shown that the *Lactobcillus* strains B54, B62, B64, and B90, which have been isolated from wheat flour by the Kazakh Scientific Research Institute of Food Industry, effectively inhibited the growth of *E. coli* and *Salmonella typhimurium.* In addition, consortium of these *Lactobacillus* strains showed a cumulative inhibitory effect.

For stabilization of antagonistic properties of a preparation on the basis of consortium duration of process of a fermentation should make not less than 48 hours and achievement of relation of cultures 1:1:1:1.

Lactobacillus may exert their antibacterial activity through the production of lactic acid and other metabolites such as hydrogen peroxide and short chain fatty acids. Also specific antibacterial compounds such as antibiotics or bacteriocins have been identified in the culture medium of several lactic acid producing bacteria. The mechanism(s) by which our strains inhibit pathogen growth is not fully understood at present. Consequently, the inhibition of the pathogen could not be ascribed simply to acidification of the culture medium due to the production of lactic acid. Proteolysis showed that compounds, which are responsible for inhibition, are proteinaceous. Furthermore, inhibition of the pathogen growth was observed at protein concentrations between 5-10 μg ml^{-1} (data not shown).

In our assay of temperature influence on inhibitory activity we establish the storage conditions for *Lactobacillus* supernatant. It is virtually stable within storage at temperature from 4 °C to 37 °C for 30 days. For maximum inhibition it is recommended that the supernatant be kept at 4 °C for 7 days. During this period inhibitory activity only decreased by 1.50-8.00% in individual strains.

Other studies, which are currently in progress, have shown that the compound(s) produced by *Lactobacillus* strains is a small (~ 3,000 Da) heat-resistant protein(s). Inhibitory activity of presented strains against pathogens could be explained by complementary action both lactic acid and protein(s). Adhesive ability of various species of lactic acid bacteria is one of obligatory criteria at selection strains for creation probioric. Sticking to epithelium cells not only prevents elimination of microbes, but also protects a mucous membrane from an opportunity of introduction of pathogenic microbes.

Veterinary tests of the preparation "Biokons" on the chickens infected Salmonella typhimurium, have shown efficiency of its use in treatment of salmonelloses, have allowed to establish the unitary medical doze of a preparation (1 ml with titer of cells of consortium 10^7) and duration of treatment (5-7 days). Introduction of probiotic to chickens daily within the first 10 days of a life in the certain doze on a background of infection with a salmonelloses and colibacterioses results to 100 % to medical effect. The lactic bacteria included in a preparation, are not pathogenic, possess adhesive properties, positively influence on normalflora gastroenteric path of chickens and promote mastering of nutrients, raising a daily average gain of alive weight of a bird. Thus, the new antimicrobic product - probiotic, capable to supervise of pathogens gastric path of bird is developed.

REFERENCES

[1] Elmer, G.W., Surawicz, C.M., McFarland, L.V. (1996) *J. Am. Med. Assoc.* 275: 870-876.

[2] Wuderlich, P.F., Braun L., Fumagalli, I., D'Apuzzo, V., Heim, F., Karly, M., Lodi, R., Politta, G., Vonbank, F., Zeltner, L. (1989) Double-blind report on the efficacy of lactic acid-producing enterococcus SF68 in the prevention of antibiotic-associated diarrhoea and in the treatment of acute diarrhorea. *J. Int. Med. Res.* 17: 333-338.

[3] Bin, L.X. (1995) Controlled clinical trial of lacteol fort sachets versus furazolidone or berberine in the treatment of acute diarrhea in children. *Ann. Pédiatr. (Paris).* 42: 396-401.

[4] Boulloche, J., Mouterde, O., Mallet, E., (1994) Management of acute diarrhea in infants and young children. *Ann. Pédiatr.* 41: 1-7 (translation and extract).

[5] Biller, J.A., Katz, A.J., Flores, A.F., Buie, T.M., Gorbach, S.L. (1995) Treatment of recurrent *Clostridium difficile* colitis with *Lactobacillus* GG. *J. Pediatr. Gastroenterol. Nutr.* 21: 224-226.

[6] Raza, S., Graham, S.M., Allen, S.J., Sultana S., Cuevas, L., Hart, C.A. (1995) *Lactobacillus* GG promotes recovery from acute nonbloody diarrhea in Pakistan. *Infect. Dis. J.* 14: 107-111.

[7] Saavedra, J.M., Bauman, N.A., Oung, I., Peerman J.A., Yolken, R.H. (1994) Feeding of *Bifidobacterium bifidum* and *Streptococcus thermophilus* to infants in hospital for prevention of diarrhea and shedding of rotavirus. *Lancet.* 344: 1046-1049.

[8] Gibson S.A.W., Conway P.L. Recovery of Probiotic Organism from Human Faeces after Oral Dosing (1994). In: *Human Health: The contribution of microorganism*, pp. 119-145. London.

[9] De Vuyst, L. and Vandamme, E.J. (1994) Antimicrobial potential of lactic acid bacteria. In: *Bacteriocins of Lactic Acid Bacteria* (De Vuyst, L. and Vandamme, E.J. Eds.), pp. 91-142. Blackie Academic and Professional, London.

[10] Tulemisova Z.K., Dudikova G.N.(2001) Efficiency new probiotic at feeding a bird. In: *Problems of scientific maintenance of agriculture* RK, Siberia and Mongolia. Almaty, 337.

[11] Diep, D.B., Håvarstein, L.S., Nes, I.F. (1995) A bacteriocin-like peptide induces bacteriocin synthesis in *Lactobacillus plantarum* C11. *Mol. Microbiol.* 18: 631-639.

[12] Tsurkan, L. (2000) Ph.D. Thesis *Lactic Acid bacteria Consortium to Control Pathogens Causing Diseases of the Digestive Tract of Animals and Poultry.*

[13] Kovalenko N.K., Kasumova S.A., Golovach T.N. (1990) Adhesion of lactic acid bacteria to intestines epithelium of agricultural animals. *Microbiological magazine* 3: 52.

In: Biotechnology in Agriculture and the Food Industry
Editor: G. E. Zaikov, pp. 35-41
ISBN 1-59454-119-1

Chapter 5

THE CO-CULTURE OF WHEAT CALLUS AND BUNT PATHOGEN AS A SUIT TEST-SYSTEM FOR THE SEARCH OF PLANT RESISTANCE INDUCERS

Igor V. Maksimov, Natalia B. Troshina, Lubov G. Yarullina, Olga B. Surina, Zalifa R. Yusupova and Ekaterina A. Cherepanova

The Institute of Biochemistry and Genetics of Ufa Research Center, Russian Acad. of Science, 450054, Ufa, pr. Oktyabrya, 69, Russia,
E-mail: phyto@anrb.ru

ABSTRACT

We have shown the possibility of prolonged co-cultivation of wheat callus (Triticum aestivum L.) and bunt pathogen (Tilletia caries (DC.) Tul. It transpired that in co-cultures the wheat calluses and bunt pathogen T. caries passes a cycle of development from teliospores germination to formation of new teliospores, virulent to common wheat. Salicylic acid (SA) influence on morphological pathogen characteristics and on protective callus cell response, connected with hydrogen peroxide production, has been studied. During the cultivation of infected wheat calluses on the medium with MS containing SA, the share of diaminobenzidine colored cells (DAB cells) in fungus growth area increased by 30%, which probably, together with the violation of the pathogen morphology caused deceleration of fungus growth. Thus, the presence of DAB-material in infected cells and the influence of SA show that their protective effect is connected with intensification H2O2 production. The obtained data on oxalate oxidase activity in wheat calluses under the influence of T. caires infection and processing with SA proves likeness to defense reactions of plant cell in vivo and in vitro. The data also disclosed one of the protective mechanisms of a well-known plant resistance inductor – SA.

Key words: wheat callus, bunt pathogen, co-culture, salicylic acid, oxalate oxidase, H_2O_2 production, DAB –cells.

The usage of traditional chemical means of plant protection is often connected with environment pollution, which requires the search of ecologically safe pesticides and ways of combating phytopathogenetic microorganisms. Primary screening of material in field conditions appeared to be the most difficult part of the work. In laboratory only fungicidity of the chemical can be examined, whereas the study of its stimulating the resistance against phytopathogens qualities is limited.

The smut and bunt pathogens considerably decrease grain-crops productivity. It is known that representatives of this fungus group develop near the crops apical meristem. Such quality of the smuts and bunts makes its possible to suggest that callus cultures can serve as an appropriate system for the study of wheat and bunt pathogen *Tilletia caries* interaction.

The increase of reactive oxygen intermediates (ROI) production as well as the increase of hydrogen peroxide (H_2O_2) production can be referred to universal protective reactions of plants to against different stressors. The basic enzyme, which is part of ROI formation, is NADP-oxidase (1). However, H_2O_2 accumulation in plant tissues may come as a result of oxalic acid oxidation by oxalate oxidase.

Lately, researchers' attention has been drawn to the possibility of salicylic acid (SA) participation in defense reactions of plant cells in various stressful conditions. Thus, for example, pathogen influence on plants leads to SA excretion into apoplast and further into the plant (2). The SA can stimulate systemic acquired resistance of plants to various pathogens (3).

Present investigation is devoted to the study of cytological characteristics of the common culture of wheat callus and bunt pathogen and to the determination of the possibility of its usage for the analysis of systemic resistance inductors influence on the morphological characteristics of the pathogen and engaging cell defense response, connected with ROI production.

Materials and Methods

Plant Material and Growth Condition

Ripe germs of common wheat *T. aestivum* were used as explants for obtaining callus material. Germs were isolated from seed-vessels and were planted onto the nutrient mediums Murashige and Scoog (MS) (4). Before the experiment calluses were cultivated at 26° C in the dark MS nutrient medium without or with addition of 0,05mM salicylic acid (SA). 10 days after the second subculture, part of the calluses were infected by bunt pathogen *Tilletia caries* (DC.)Tul teliospores, reproduced on hothouse common wheat plants. About 80-100 spores constituted *infection load* per each callus. 20-30 days later common cultures of wheat bunt *T. caries* and wheat were grafted onto the fresh nutrient mediums.

The Cytological Detection of H_2O_2 Production

The cytological detection the hydrogen peroxide generation with oxalate oxidase participation in oxidation the chromogenic substratum 3,3 – diaminobenzidine (DAB) (5). After DAB-coloring calluses were fixed in ethanol and acetic acid (3:1).

The Biochemical Detection of Oxalate Oxidase Activity

For biochemical assessment of oxalateoxidase calluses were homogenized in phosphate buffer pH 6,0 (1:3). Homogenate was centrifuged (3000 rotation per minute, 20 min.) and oxalate oxidase activity was detected in cytoplasmic fraction. After multiple washing away cytoplasmic proteins, the sediment on cell walls was 1M NaCl treated for educing ion binding proteins with cell wall oxalate oxidase.

The activity of oxalate oxidase in the fractions under analysis was determined with the help of micro method (6). The micro titration plate was scanned by 492 nm wave on photometer "Uniscan" (Great Britain). Protein contents were determined by means of Bradford method (Bradford, 1976). Uninfected wheat calluses growing on nutrient mediums MS without SA, were used for control. The experiments were carried out 3 times and 3-4 calluses were fixed in each variant.

Results

Visual observations of mycelium *T. caries* growth on wheat callus have shown that white and downy fungus mycelium could be seen on the 5th day after inoculation. Callus infections resulted in the increase of the culture's crude weight (fig. 1), as well as in the increase the number of mitoses and the size of cells. We could also observe gradual growth of both mycelium mass on callus surface and of the callus mass throughout 3 months of the experiment.

Cytological analysis of the uninfected calluses has shown that after the teliospores germination fungus mycelium was developing both inside the callus and in hypha, which were mostly sited in intercellular space (fig. 2a). From the 15th day the fungi hypha occupied callus cells (fig. 2b). Hyphas were of two forms: stick-like (were strongly basophilic), and beads-like (slightly colored by basic dyes) (fig. 2c).

30 days after inoculation fungus mycelium gradually formed teliospores on callus surfaces (fig. 2b), and after 45 days it formed soruses (fig. 2c). The color and the size of the formed teliospores did not differ from the size and the color of those, used for inoculation.

Other data were obtained on the common culture of wheat callus and *Neovossia indica*. It appeared that within 60 days of cultivation on wheat calluses, the pathogen hypha grew only on the surface of callus. They formed teliospores which did not reach the size of the teliospores used for inoculation, i.e. they remained unripe within 10 months (8).

It was of interest for us to find out whether bunts pathogen teliospores, which were formed on calluses, could infect wheat. Teliospores were carefully taken from calluses which had been cultivated for a long time with bunt pathogen. Afterwards, wheat seeds of a sensitive to the pathogen Znitsa variety were inoculated with these teliospores. 17% of infected ears were obtained in conditions of a hothouse.

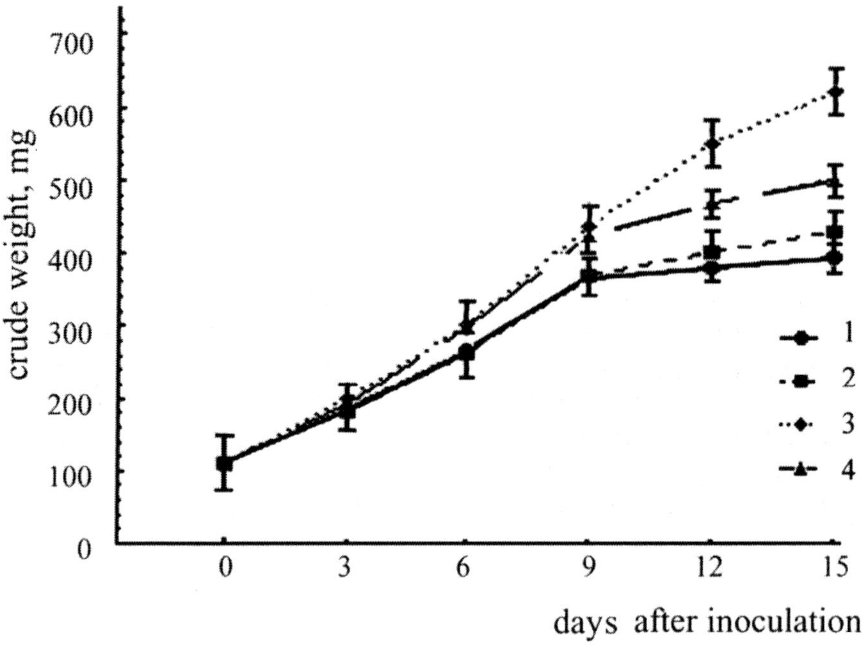

Fig. 1. *T. caries* growth on wheat calluses after processing with salicylic acid (SA). 1- control (MS), 2- MS + 0,05 mM SA, 3- *T. caries* (MS), 4 - *T. caries* (MS + 0,05 mM SA).

Thus, we have shown a possibility of prolonged co-cultivation of wheat callus and bunt pathogen. It transpired that in co-cultures the wheat calluses and bunt pathogen *T. caries* passes a cycle of development from teliospores germination to formation of new teliospores, virulent to common wheat.

SA influence on morphological pathogen characteristics and on protective callus cell response, connected with hydrogen peroxide production, has been studied on the co-culture of wheat callus and bunt pathogen.

On wheat calluses cultivated on nutrient medium MS containing SA, suppression of surface pathogen growth and formation of necrotic zones in the sites of fungus mycelium growth were observed within the first 5-10 days after the germination of teliospores (fig. 2). Such phenomenon was observed by us for the first time, and probably it characterizes the protective callus cell response like the response of resistant plants to the infection by an incompatible form of phytopathogens (9). Under the influence of SA, the ratio of fungi stick-like and beads-like hypha: the share of stick-like decreased in comparison with beads-like (fig. 2d).

In uninfected calluses and in calluses growing on the medium containing SA, DAB coloring was typical only of rhizoid cells. In the process of infecting, parenchyma-like cells (about 20%) with DAB –colored material were noticed in the fungus penetration zone. In the cytoplasm of some cells there were fungus hyphas, which were also colored by DAB method.

During the cultivation of infected wheat calluses on the medium with MS containing SA, the share of DAB cells in fungus growth area increased by 30%, which probably, together with the violation of the pathogen morphology caused deceleration of fungus growth.

As the experiments have shown, on the 10th day after callus infection, oxalate oxidase activity in the cytoplasmic fraction of wheat cells increased after callus inoculation the fungi as well as after cultivating on the nutrient medium_containing SA (table 1). The highest oxalate oxidase activity was observed in the calluses which grew on medium containing SA and infecting by bunt pathogen, which proves the engagement of this enzyme in protective reactions.

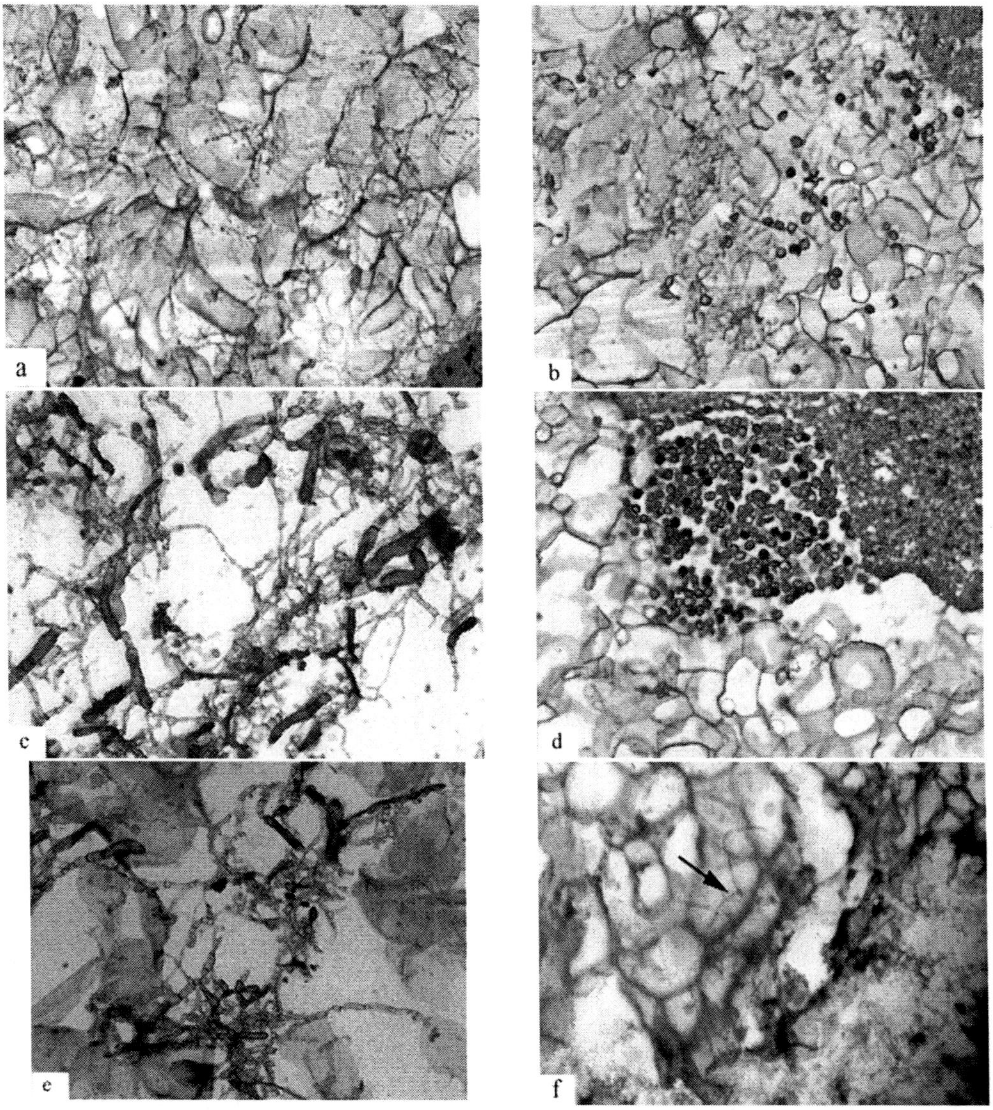

Fig. 2. The stages of *T. caries* growth and development on wheat calluses: a) intercellular mycelium, x 600; b) intracellular mycelium and the beginning of spores formation intercellular areas, x 600; c) stick-like and beads-like fungi hypes, x 900; d) the fungi soruses formation, x 600; e) the change of ratio of various types of hyphes after processing with SA, x 900; f) DAB material in the cells of calluses. Arrows show cells with pathogen hyphes, x 800; a -15, b- 30, c-45, d, e -20 and f-10 days after inoculation

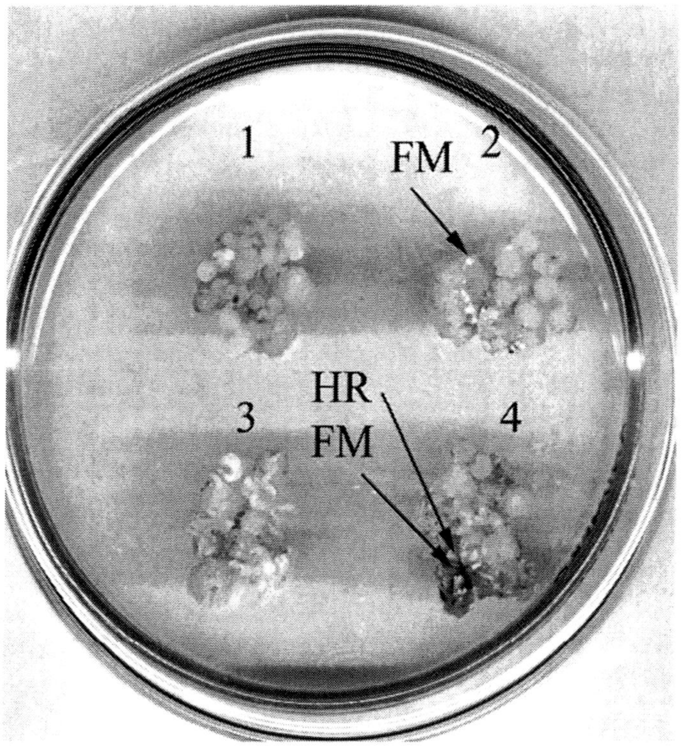

Fig. 3. Influence of salicylic acid on the co-culture of wheat callus and *T. caries* (fungal mycelium (FM)). The necrotic zones (HR) to formation of in mycelium growth sites (follow the arrows). 1- control (MS), 2- *T. caries* (MS), 3- MS + 0,05 mM SA, 4 - *T. caries* (MS + 0,05 mM SA).

Table 1. The influence of co-cultivating the wheat calluses and the bunt pathogen *T. caries* on the MS medium with or without salicylic acid on oxalate oxidase activity

Variant	Oxalate oxidase activity (activity unit on grams of crude weight)	
	Cytoplasmatic fraction	Extracellular fraction
Control calluses	7.1±0.5	0.23±0.02
Calluses inoculated by bunt pathogen	17.0±1.1	0.17±0.01
Calluses growing on salicylic acid	19.1±0.9	0.25±0.03
Calluses growing on salicylic acid and inoculated by bunt pathogen	24.2±1.2	0.29±0.02

The analysis of extra cellular fraction revealed that under the influence of infection, inhibition of this ferment excretion into intercellular space takes place. SA restored callus cell capability to the secretion of this ferment.

Thus, the presence of DAB-material in infected cells and the influence of SA show that their protective effect is connected with H_2O_2 production intensification. There are a lot of data on oxalate oxidase participation in protective reactions of plants, during infection by phytopathogen (10]. Activity expression of these proteins in plants under the influence of phytopathogens and their elicitors has been discovered (11). The obtained data on oxalate oxidase activity in wheat calluses under the influence of *T. caires* infection and processing with SA, prove likeness to defense reactions of plant cell *in vivo* and *in vitro*. The data also disclosed one of the protective mechanisms of a well-known plant resistance inductor – SA.

Thus, the co-culture of wheat calluses and biotrophy pathogenic fungus *T. caries* is a convenient model for visual assessment of fungus growth and development. The model can be used for the study of various plant resistance inductors' protective mechanisms.

ACKNOWLEDGEMENTS

This work has been fulfilled thanks to financial support of PFBR (project #02-04-97923)

REFERENCES

[1] R. Mittler: *Trends in Plant Science*. **7**(9), 405 (2002).

[2] N.B. Vasyukova, S.V. Zinovjeva, N.G. Iljinskaya *Dokl. Biol Science (Russian)*, 364, 843 (1999).

[3] Z. Chen, H. Silva, D.F. Klessig: *Science*. 263. 1883 (1993).

[4] T. Murashige, F.A. Scoog: *Physiol. Plant*. 15 (3), 473 (1962).

[5] H. Thordal-Christensen, Z. Zhang, V. Wei, D.B. Collinge: *Plant J*. 11, 1187 (1997).

[6] R.M. Khairullin, L.G. Yarullina, N.B. Troshina, I.A. Akhmetova: *Biochemistry (Russian)*, 66(3), 354 (2001).

[7] M.M. Bradford: *Anal. Biochem*. 72, 248 (1976)

[8] S. Kaur, A. Malhotra, S.S. Aujla, K.S. Gill: *Ind J. Exp. Biol*. 2, 193 (1990)

[9] M.E. Alvarez: *Plant Molecular Biology*. 44, 429 (2000)

[10] B. Dumas, G. Freyssinet, R.E. Pallet: *Plant Physiol*. 107(4) 1091 (1995)

[11] Y. Wei, Z. Zhang, C.H. Andersen, E. Schmelzer, P.L. Gregersen, D.V. Collinge, V. Smedegaard-Petersen, H. Thordal-Christensen: *Plant Molecular Biology* 36, 101 (1998)

In: Biotechnology in Agriculture and the Food Industry ISBN 1-59454-119-1
Editor: G. E. Zaikov, pp. 43-48

Chapter 6

Preparation of Succinimidyl Active Esters of N,N'-Tetrametil-5(6)-CarboxyRhodamine and 5(6)-Carboxy-X-Rhodamine

Goodoff P. Vladimir, Alekseev I. Yakov and Lunin G. Vladimir
All-Russia Research Institute of Agricultural Biotechnology (ARRIAB)
Timiryazevskaja street, 42, Moscow, 127550, Russia.
Phone/fax: (095) 977-74-55
E-mail: goodoff@hotbox.ru
www.agrobiotech.ru

Abstract

A mixture of 5- and 6-isomers of N,N'-Tetrametil-5(6)-CarboxyRhodamine and 5(6)-Carboxy-X-Rhodamine was activated with 1-[3-(dimethylamino)propyl]-3-ethylcarbodiimide hydrochloride in the presence of either N-hydroxysuccinimide to give the corresponding succinimidyl esters. The regioisomeric mixtures were separated to give 5- and 6-succinimidyl active esters in >98% purity.

Keywords: Succinimidyl, active esters, rhodamine dyes, fluorescent dyes

Introduction

Rhodamine derivatives have been widely used as fluorescent markers in bioanalytical chemistry (1-5) and have found exceptional utility as the fluorophores of choice in fluorescent polarization immunoassays (FPIA) used for monitoring the level of drugs and hormones in human fluids (6,7). Active esters of N,N'-Tetrametil-5(6)-CarboxyRhodamine and 5(6)-Carboxy-X-Rhodamine are used for production of labels and conjugates (1,4,8-10). From the literature (13,14) only limited information is available for the preparation, purification, and analysis of the N,N'-Tetrametil-5(6)-CarboxyRhodamine and 5(6)-Carboxy-X-Rhodamine succinimidyl active esters, which usually possess higher hydrolytic stability (15), have not been previously reported. Thus, there is a need for an efficient and cost effective method for the preparation of these active esters. We now report a method for the preparation N,N'-Tetrametil-5(6)-CarboxyRhodamine and 5(6)-Carboxy-X-Rhodamine succinimidyl (2a, 2b, 2c, 2d) active esters from the inexpensive, commercially available N,N'-Tetrametil-5(6)-CarboxyRhodamine (1a) and 5(6)-Carboxy-X-Rhodamine (1b) as shown in Scheme 1.

Scheme 1

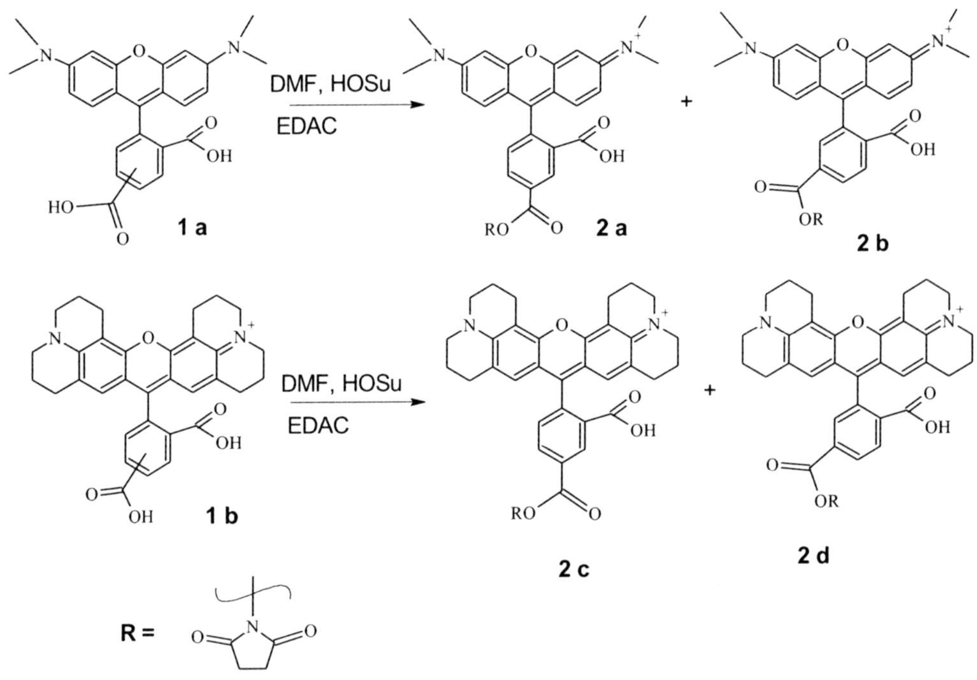

MATERIALS AND METHODS

N '-Tetrametil-5(6)-CarboxyRhodamine and 5(6)-Carboxy-X-Rhodamine were obtained on standart method (13) from compaunds, while purchased from Aldrich Chemical Co. (Milwaukee, WI). Solvents were of HPLC grade and used without further purification. Silica gel 60 (230-400 mesh) was purchased from Merck (Germany). FPLC purification system and column HR30 (1,6×40 cm) were purchased from Pharmacia (Sweden). ^1H NMR spectra were recorded at 300 MHz, on a Brucker spectrometer. Electrospray ionization mass spectra were recorded on a Perkin-Elmer Sciex API 100 instrument. Analytical HPLC was performed using a Gilson HPLC system (France), C_{18} column 8×100 mm, a flow rate of 1 ml/min and UV detection at 260 nm.

N,N'-Tetrametil-5(6)-CarboxyRhodamine succinimidyl active esters (2a, 2b)

To a solution of N,N'-Tetrametil-5(6)-CarboxyRhodamine (5.0 g 11,6 mmol) in anhydrous N,N'-Dimethylformamide (DMF: 50 ml) was added 1-[3-(dimethylamino)propyl]-

3-ethylcarbodiimide hydrochloride (EDAC: 3,1 g, 16 mmol) followed by N-hydroxysuccinimide (HOSu: 1,9 g, 16 mmol). The reaction was covered with foil, stirred under argon, and monitored by analytical HPLC (CH_3CN gradient (5% - 80 % for 60 min) in H_20 TEAA buffer pH=6,5). After 4,5 h, additional EDAC (510 mg, 0,2 eq.) was added, and the reaction was stirred for an additional 18,5 h. The reaction mixture was rinsed into a separatory funnel with DMF (10 ml) and diluted with acetone (200 ml). Buffer (0.05 M, pH=6 phosphate buffer, 250 ml) was added, and the mixture extracted with diethyl ether (Et_2O)/ ethyl acetate (EtOAc) (1:1, 300 ml). The organic layer was separated, and the aqueous layer was extracted two times with Et_2O/ EtOAc (1:1, 250 ml). The combined organic extracts were washed with water (3×100 ml) and brine (1×250 ml), dried over Na_2SO_4, and filtered, and the solvents were removed *in vacuo* to afford 4,5 g. of crude N,N'-Tetrametil-5(6)-CarboxyRhodamine succinimidyl active esters. Separations of the isomers on the FPLC system was performed as follows: (a) the HR30 column was equilibrated with 2L of eluent (85% acetone/ 15% toluene); (b) crude active ester mixture, adsorbed onto silica gel (50 g.), was loaded into the SIM (Stainless steel sample injection module); and (c) the isomers were eluted at a mine system pressure of 100 psi and solvent pressure of 60 psi, to afford 1,4 g. of 2a (< 2% of 2b present) and 1,1 g. of 2b (< 3% of 2a present). For further purification, each lot was adsorbed onto silica gel (12 g.), loaded into the SIM, and purified using the previous conditions to afford 795 mg of 2a (20% yield < 0.1% of 2b present) and 775 mg of 2b (22% yield < 0.1% of 2a present). Complete spectral and analytical data are shown in Tables 1 and 2.

Table 1. Preparation of Succinimidyl Active Esters of N,N'-Tetrametil-5(6)-CarboxyRhodamine and 5(6)-Carboxy-X-Rhodamine 2 a-d.

Compd	Yield (%)	HPLC		Molecular formula	ESI – MS $(M+H)^+$
		Retention time (min)	Purity (%)		
2 a	20	37.43	99	$C_{29}H_{26}N_3O_7$	528.5
2 b	22	34.65	98	$C_{29}H_{26}N_3O_7$	528.5
2 c	18	33.25	99	$C_{37}H_{34}N_3O_7$	632.6
2 d	19	30.76	97	$C_{37}H_{34}N_3O_7$	632.6

Table 2. NMR Data for Active Esters 2 a-d.

Compd	^{1}H NMR (DMSO – d_6) (δ, PPM; J, Hz)
2 a	12.30 (s, 1H), 8.66 (d, 1H, J=1.96), 8.35 (d, 1H, J=8.15), 7.82 (s, 1H), 7,41 (s,H1), 6,96 – 6,85 (m, 4H), 6.28 –6,15 (m, 4H), 3.55 (s, 3H), 3,54 (s, 3H), 3 (s, 6H), 2, 99 (m, 4H)
2 b	12.30 (s, 1H), 8.66 (d, 1H, J=8.15), 8.35 (d, 1H, J=1.9), 7.82 (s, 1H), 7,41 (s,H1), 6,96 – 6,85 (m, 4H), 6.28 –6,15 (m, 4H), 3.55 (s, 3H), 3,54 (s, 3H), 3 (s, 6H), 2, 99 (m, 4H)
2 c	12.30 (s, 1H), 8.66 (d, 1H, J=1.96), 8.32 (d, 1H, J=8,25), 7.82 (s, 1H), 5.99 (s, 1H), 4.09 – 3.86 (m, 4H), 3.00 – 2.86 (m, 4H), 1.97- 1,66 (m, 8H)
2 d	12.30 (s, 1H), 8.66 (d, 1H, J=8.15), 8.32 (d, 1H, J=1,9), 7.82 (s, 1H), 5.99 (s, 1H), 4.09 – 3.86 (m, 4H), 3.00 – 2.86 (m, 4H), 1.97- 1,66 (m, 8H)

Flash Chromatography

The crude reaction mixture (4,5 g) adsorbed onto silica gel (50 g.) was loaded onto an equilibrated flash column (14× 30 cm bed of silica) and separated (85% acetone/ 15% toluene) to afford 1 g. of 2a (< 0.1% of 2b present) and 950 mg. of 2b (15 % of 2b present).

5(6)-Carboxy-X-Rhodamine Active Esters (2c, 2d)

To a solution of 5(6)-Carboxy-X-Rhodamine (5.0 g 9,3 mmol) in anhydrous N,N'-Dimethylformamide (DMF: 50 ml) was added 1-[3-(dimethylamino)propyl]-3-ethylcarbodiimide hydrochloride (EDAC: 2,7 g, 14 mmol) followed by N-hydroxysuccinimide (HOSu: 1,63 g, 14 mmol). The reaction flask was covererd with foil, and the reaction was monitored by analytical HPLC (CH_3CN gradient (5% - 80 % for 60 min) in H_20 TEAA buffer pH=6,5). After stirring under argon for 25 h. The reaction mixture was rinsed into a separatory funnel with DMF (10 ml). Buffer (0.05 M, pH=6 phosphate buffer, 250 ml) was added, and the mixture extracted with chloroform (4 times, 200 ml). The organic layer was separated. The combined organic extracts were washed with water (3×100 ml) and brine (1×250 ml), dried over Na_2SO_4, and filtered, and the solvents were removed *in vacuo* to afford 4,1 g. of crude 5(6)-Carboxy-X-Rhodamine active esters. Separations of the isomers on the FPLC system was performed as follows: (a) the HR30 column was equilibrated with 2L of eluent (92 %acetonitrile/ 8% toluene); (b) crude active ester mixture, adsorbed onto silica gel (25 g.), was loaded into the SIM; and (c) the isomers were eluted at a mine system pressure of 100 psi and solvent pressure of 60 psi, to afford 1 g. of 2c (18% yield < 1% of 2d present) and 1,1 g. of 2d (19 % yield < 1% of 2c present).

Flash Chromatography

The crude reaction mixture (4,1 g) adsorbed onto silica gel (50 g.) was loaded onto an equilibrated flash column (14× 30 cm bed of silica) and separated (85% acetone/ 15% toluene) to afford 950 g. of 2c (< 0.1% of 2d present) and 1,2 g. of 2d (15 % of 2c present). Complete spectral and analytical data are shown in Tables 1 and 2.

RESULTS AND DISCUSSION

The mixture of 1a and 2b was reacted with EDAC and HOSu in anhydrous DMF to convert each acids to its respective succinimidyl active esters (2a, 2b, 2c, 2d). Workup conditions consisted of pouring the crude reaction mixture into a separatory funnel, adding pH 6 buffer solution, end extracting with a mixture of ether/ethyl acetate (1;1) for 1a, or chloroform for 1b. The selection of solvents combination was necessary to fully maximize extraction of the desired active esters and minimize the presence of HOSu, while the choice of pH 6 phosphate buffer prevented undesired hydrolysis. This workup active esters - 2a, 2b, 2c and 2d with <1 % of the starting acids. Separation of the active esters was attempted according to two different methods: flash chromatography and FPLC purity system.

Separation by flash chromatography was inefficient since it did not afford good resolution between the two rhodamine succinimidyl esters regioisomers. A second method was sought that would combine the ease of use of flash chromatography with separation capabilities of HPLC. The FPLC system was attractive it is designed for milligram to kilogram scale separations and provides solutions to critical operational problems typically found with glass columns: speed, safety, and performance. For gram scale separations, column operate at higher pressures (85-120 psi) with higher flow rates (100 – 300 ml/min) than flash chromatography, which results in shorter run times. This self-contained system reduced the concern of using glass columns under pressure, while its radial compression technology eliminates voids and channels that often occur in conventional flash chromatography.

Separation of the crude mixture of rhodamine succinimidyl esters using FPLC system afforded good separation of 2a from 2b and 2c from 2d (both >97% by analytical HPLC). Further purification afforded 2a, 2b, 2c and 2d with purity > 99%, see Table 1. A disadvantage for any silica gel based chromatographic separation of active esters is hydrolysis of the active ester while on the column, which frequently leads to diminished product yields. This method of separation was found to be the most effective for the production of active esters.

In summary, our method transforms the inexpensive, commercially available N,N'-Tetrametil-5(6)-CarboxyRhodamine and 5(6)-Carboxy-X-Rhodamine to the highly useful succinimidyl active esters (2a, 2b, 2c and 2d). Separations on the FPLC system gave very pure (>99%) succinimidyl active esters. These procedures allow for the rapid and cost effective production of these carboxyrhodamine active esters, and the FPLC system offers the advantages of safety, reproducibility, and speed over other techniques for the separations of larger amounts of material.

REFERENCES

[1] Haughland, R.P., D. L. Wise and D. Wingard, eds. *Biosensors with Fiberoptics*, p. 85 – 109, Humana Press, Clifton, NJ. (1991)

[2] Bright, F.V. Bioanalytical Applications of Fluorescence Spectroscopy. *Anal. Chem.* 60, 1031A-1039A. (1988)

[3] Owens, M.A., Loken, M.R. *Flow Cytometry: Principles for Clinical Laboratory Practice,* pp. 28-29, Wiley-Liss, New York. (1995)

[4] Haughland, R.P., *Handbook of Fluorescent Probes and Research Chemicals*. Molecular Probes, Eugene, OR. (1992)

[5] Miki, M. The Recovery of the Polymerizability of Lys-61-Labelled Actin by the Addition of Phalloidin. *Eur. J. Biochem.* 164, pp. 229-235. (1987)

[6] Lu-Steffes, M., et al. Fluorescence Polarization Immunoassay 4. Determination of Phenitoin and Phenobarbital in Human Serum and Plasma. *Clin. Chem.*28, pp. 2278-2282.

[7] Adamczyk, M., Harington, C., and Jhonson, D. Reagents and Methods for the Quantification of Imipramine and Desipramine in Biological Fluids. *U.S. Pat. 5340750.* (1994)

[8] Bodanszky, M. *Principles of Peptide Synthesis*, pp. 28-49, Springer-Verlag, New York.

[9] Bodanszky, M. *The Peptides*, Vol. 1, pp. 106-175, Academic Press, New York. (1979)

[10] Reference 4, p. 26.

[11] Yoshida, R.A. Support Ligand Analog Fluorescent Conjugate and Serum Assay Method Involving Such Conjugate. *Eur. Pat. Appl. 0015695.* (1980)

[12] Vanderbilt, A.S., Osikowicz, E.W., Fino, J.R., Shipchandler, M.T. Total Estriol Fluorescence Polarization Immunoassay. *Eur. Pat. Appl. 0200960.* (1986)

[13] L. G. Lee, S. L. Spurgeon, C. R. Heiner, S. C. Benson, B. B. Rosenblum, S. M. Menchen, R. J. Graham, A. Contantinescu, K. G. Upadhya and J. M. Casel New energy transfer dyes for DNA sequencing, 2816–2822 *Nucleic Acids Research*, Vol. 25, No. 14 (1997)

In: Biotechnology in Agriculture and the Food Industry ISBN 1-59454-119-1
Editor: G. E. Zaikov, pp. 49-50

Chapter 7

ECOBIOTECHNOLOGY TO PRODUCE LAWN GRASS AS A TOOL AGAINST HIGH CONCENTRATIONS OF COPPER ION AND SODIUM CHLORIDE

***E. A. Gladkov*[1]*, Yu. I. Dolgikh*[2] *and V. V. Biryukov*[1]**
[1]Moscow State University of Engineering Ecology
[2]Timiryazev Institute of Plant physiology RAS
fax: 977 80 18, e-mail: gsc@ippras.ru

Soils pollution produced by industry, transport and application of antiicing reagents considerably inhibite the growth of lawns in large cities[5]. Grass yield losses due to these stress factors can be reduced by using resistant cultivars.

Heavy – metals are the most toxic elements among the chemical ones. Only penticides are more poisonous than heavy – metals. In the most cases these elements can be arranged as: Cu>Ni>Cd>Zn>Pb>Hg, according to their toxicity.

Heavy – metal pollution can result in some serious physiological and anatomical disturbances in plant grouth and development. Town lawns especially acutely suffer from copper salt pollution.One of the methods to solve this problem is to produce some plant lines, resistant heavy metals.

Application of the biotechnological approaches could accelerate and facilitate the method of conventional breeding[1, 2, 3, 4, 6]. To select tolerant plants the sensitivity of *Festuca rubra* and *Agrostis stolonifera* to NaCl, $CuSO_4$ and $CdCl_2$ have been tested both state in intact plants and cultivated tissue.

The rate of seed germination and dry weight per plant on the 30 d of growth in the soil, supplemented with 3-10 g/kg of soil NaCl, 75-300 mg/kg $CuSO_4$ or 2-6 mg/kg $CdCl_2$ were determined. These concentrations of salts corresponded to the real levels of pollution in Moscow.

The germination of seeds was not essentially affected by Cd. NaCl concentration 3 g/kg and more inhibited seed germination and plant growth, 10 g/kg NaCl concentration was sublethal for both grasses. *A.stolonifera* proved to be relatively tolerant to Cu up to 150 mg/kg. Toxic effect of Cu on *F. rubra* was revealed at 75 mg/kg. Both grass got yellow when 300 mg/kg Cu was applied.

Embryogenic callus of F. rubra and A. stolonifera have been produced from seeds in MS medium supplemented with 3 mg/l 2,4-D. To regenerate the plants, the calli were transfered to hormon-free MS medium. Later NAA (2 mg/l) was added to enhance root formation. To isolate resistant lines callus tissues were cultivated in the media with NaCl in selective concentration 10 and 20 mg/l. Cultivated cells of *F. rubra* and the intact plants were shown to be more resistant with to NaCl than callus and plants of *A. stolonifera*. The callus of *A. stolonifera* was only able to regenerate some shoots in the medium containing 10 mg/l NaCl and to form roots in the medium without salt. The plants of *F. rubra* and *A. stolonifera* were regenerated in the medium with 20 mg/l NaCl and they were rooted in the medium with 10 mg/l NaCl. This resistance to sodium chloride in herited in the following generation.

To select some cells resistant high copper ion concentration, we cultivated calli in the MS medium supplemented with sublethal copper doses of 150 mg/l during two subculturings. Shoots were regenerated in the medium supplemented with 100 mg/l copper. As a result, red fescue callus turned out to be more susceptible then fine bent ones.

On the whole, 56 fine bent and 10 red fescue regenerants were selected. Red fescue regenerant died a month after their planting in the soil. The most part of fine bent regenerants had normal morphology and good growth. Five of them were in flour and produced some seeds. The regenerants, growing in the soil with 150 mg/l copper, didn't differ from the plants growing in the soil without copper supplement. The initial plants delayed their growth and got yellow at this copper concentration.

To find out this trait inheritance we sprouted the seeds of one regenerant in the soil supplemented with 150 mg/l copper. This regenerant seed germination has three times as high as it was in initial plants. Thus, the most fine bent regenerants produced from copper resistant cell lines, were copper-ion tolerant. This resistance is herited in the following generation.

References

[1] Domozlicka E., Opatrny Z. The effect of cadmium on tobacco cell culture and the sebction of potentially Cd-resistant cell lines. // *Biol. Plant.*, 31(1), 19-27, 1989

[2] Dolgikh Yu.I. High level of variability among the plants regenerated from callus of inbred A188 Maize *Genetics Cooperation Newsletter*. 1999. P. 70-71

[3] Gilissen L.J.W., Staveren M. J. Zinc-resistant cell lines of Haplopapput gracilis// *J.plant Physiol.* v. 125, P. 95-103, 1986

[4] Glinski J., Turski R., Heavy metal pollution of polish soil//*New enviromental aspects of land use in Poland and Hungary*, 1993, №40, p. 1-5.

[5] Shevyakova N.I., Klochkova V.S., Lomova M.G., Malyuzhenets E.E., Kuznetsov Vl.V. Peculiarities of the salt stress effect on plants in urban conditions. // *Proceedings of the intern.conf. on Ecology in cities*, 1998, Greece, p.27.

[6] Smith M.K., McComb J.A. Selectionfor NaCl-tolerance in cell cultured of Medicago sativa and recovery of plant cell line// *Plant Cell Repts,* 2, 126-8, 1983

In: Biotechnology in Agriculture and the Food Industry ISBN 1-59454-119-1
Editor: G. E. Zaikov, pp. 51-61

Chapter 8

COMPLEX DIAGNOSTICS OF PHYTOPATHOGENS IN ECOLOGICAL TRIALS OF POTATO STRAINS

***Natalia A. Lykova*[1], *Tatiana A. Danilova*[2], *Svetlana M. Sinitsyna*[2], *Zinaida Z. Yevdokimova*[3], *Ludmila N. Golovina*[2], *Zinaida P. Kotova*[2], *Ivan I. Kozub*[2] and *Yelena B. Mileyeva*[3]**

[1] Agrophysical Research Institute of the Russian Academy of Agricultural Sciences, 14 Grazhdanskiy Prospekt, Saint-Petersburg, 195220, Russia; e-mail: ivl@agrophys.spb.su; fax: 7(812)-534-1900;

[2] Northwest Scientific and Methodological Center of the Russian Academy of Agricultural Sciences, 5 Shosse Podbel'skogo, Pushkin, Saint-Petersburg, 196606, Russia; e-mail: sznmc@quantum.ru; fax: 7(812)-466-6476

[3] Leningrad Research Institute of Agriculture, 1 Institutskaya ul., p/o Belogorka, Gatchinskiy District, Leningradskaya Oblast, 188388, Russia; e-mail: steff@mail.ru; fax: 7(812)-71-51132

ABSTRACT

Scientifically substantiated models of region-adapted potato strains has been developed along with an approach to ecological trials of the most promising selection products providing for express disclosure of the products best adapted to local agrolandscapes. Adaptive characteristics of novel potato strains have been studied using complex diagnostics of strain-phytopathogen interactions. Potato mycoses, bacterioses, and viroses were tested. An algorithm for express complex data treatment (rapid ecological crop trial, RECT) has been developed and used to determine promising potato strains resistant to mycoses, bacterioses and viroses.

Keywords: mycosis, bacteriosis, virosis, complex diagnostics, ecological trial of cultivars, productivity

INTRODUCTION

New Trends in Diagnostics of Phytopathogens during Ecological Trials

Promising candidate strains of potato are potential novel cultivars and so belong to the range of the most important factors of enhancement of the productivity of the crop, stabilization of the whole agricultural sector, and improvement of the final product quality. Therefore, from the very start, they require testing in different geographical zones to determine their adaptive potential, i.e., the ability to adapt to specific soil and climate conditions. A high plasticity and adaptability of potato in different geographical zones is reported (Lorkh, 1968; Beukema and Zaag, 1990). In this connection, determination of ecological responses to environment was included into the complex of agricultural and biological trials of potato strains and hybrids. The environment may be regarded as a combination hydrothermal, mineral, and phytopathogenic parameters.

The Russian gene bank of potato is esteemed as having the universal importance (Kiru, 2001). By now, potato strains tolerant to widespread diseases including those prevalent in Russian Northwest have been developed. In testing of candidate potato strains in the Northwest of Russia, the above environmental parameters were taken into account. However, there is the risk of introduction of other bacterial and fungal diseases including dry scab contaminating potato seeds exported from Holland, Sweden, and Poland, and of the spread of the diseases in the Northwest of Russia. Therefore, under development is a novel complex approach to express evaluation of candidate strains for the range of their environmental responses with regard to their resistance to phytopathogens. Such approach to studying of the susceptibility of potato plants to diseases will provide for selection of candidate strains that are tolerant to the totality of possible phytopathogens and to determine the group of phytopathogens to which a specific strain is susceptible.

In many countries old potato strains are still popular. However, marketing studies have shown a high interest of potential customers (farmers and amateur horticulturists) in novel potato strains having a high resistance to late blight, viral diseases, and tuber lesions (Swiezynski, 1987). In Europe, the late blight agent (*Phytophthora infectans*) is one of the most dangerous phytopathogens affecting potato plants (Schüber and Schiessendoppler, 1987). In regions featuring drastic weather changes potato strains are needed having a high starch content (17-22%) and, at the same time, an attractive appearance, good taste, high cooking quality, and resistance to tumor development, viral diseases, and late blight.

The aim of our study was to enhance the objectiveness of evaluation of the resistance of cultivars to phytopathogens including mycoses, bacterioses, and viroses by using a parametric approach to candidate strain trials and an integrative scheme of data treatment.

MATERIALS AND METHODS

Potato Planting

The planting arrangement was the same as in the standard scheme, i.e., one experimental plot comprised four rows of 15 tubers making 60 tubers in a plot. Four plots were used in one experiment making 60×4=240 tubers per experiment. The planting pattern was 70×35cm. Plot length, width, and area were 4.9 m, 2.8 m, and 17.72 m^2, respectively. Growing space for one plant was 0.245 m^2. The plot arrangement was randomized.

Characterization of Edaphic and Hydrothermal Conditions

The experiments were carried out at 5 sites of the Northwest Region of Russia: in Murmanskaya Oblast (A), Arkhangelskaya Oblast (B), Kaliningradskaya Oblast (C), Leningradskaya Oblast (D), and Karelia (E). The experimental period lasted from 2001 till 2003.

The climate in the Northwest Region on moderately continental, the summer is cool and short. At site A (Northern limit of cultivation of a potato), 24-h sunlight lasts for about three summer months and frosts are possible on any summer day.

To characterize humidity conditions during active vegetation (see Table 1), we calculated the Selianinov's hydrothermal coefficient (HTC) by the following formula (Losev, 1994):

$$HTC = \frac{\Sigma r}{0.1\Sigma t_{>10^0 C}}$$

where $\Sigma \boldsymbol{r}$ is total sediments during the time when air temperature was above 10°C and $\Sigma t_{>10^0 C}$ is the sum of air temperatures during this time.

Table 1. Working edaphic and hydrothermal environmental factors at trial sites.

trial sites	humus %	NPK, mg/kg	total sediments, mm	the sum of temperatures >10°C, °C	HTC	Humidity
A	3,2	234-595	159,7-226,0	924,3-1155,0	1,73-1,96	extra wet
B	2,3-5,7	64-1186	128,1-133,3	1326,9-1453,0	0,92-0,97	arid
C	4,2	144	128,3	1456,2	0,88	arid
D	3,3-3,5	312-536	154,5-264,3	1356,2-1485,7	1,00-1,95	dry - extra wet
E	1,9-2,1	148-1169	119,1-244,8	1286,7-1526,8	0,92-1,60	arid - humid

Humidity was categorized as arid, dry, humid or extra wet.

Before an experiment, soil nitrogen (N_{total}), phosphorus (P_2O_5) and potassium (K_2O) levels were determined and added to amounts introduced with fertilizers. The fertilizers were introduced at amounts (kg/ha) of N80P80K80 (Control), N120P120K120 (Background 1) or N160P160K160 (Background 2). The ranges of actual NPK amounts in the soil after the addition of fertilizers are shown in Table 1.

Organic fertilizers were introduced beforehand to make 60 tons/ha.

Potato Strains

Five potato strains were tested at different backgrounds of mineral nutrition and two different times of harvesting. The strains were Nevskiy, Yelizaveta, Charodey, Peterburgskiy, Belogorskiy.

Phytopathogens

A complex diagnostics of strain-pathogen interaction was employed. During the vegetation period, the vulnerability of planted potato to diseases was assessed visually. The resistance of the plants to late blight, black scab (*Rhizoctonia* disease), leaf spot (*Macrosporium* disease), early blight (*Alternaria* disease), and black dot (anthracnose) under conditions of natural infectious background was scored by a 9-grade scale: 1 – all leaves are lesioned; 3 – more that 50% of leaves are lesioned; 5 – 25-50% of leaves are lesioned; 7 – less than 25% of leaves are lesioned; 8 – separate lesions are present on leaves; 9 – no lesions.

Analysis of potato tubers for their lesioning by diseases was carried out after harvesting, in parallel with evaluation of crop yields and with classification of tubers by their shape. Each tuber was examined visually for disease manifestations, physiological fissures, and spindle shape. The resistance of the tubers (%) to mycoses including common scab, dry scab, late blight, skin necrosis (phomosis), dry rot, and black scab was estimated.

Bacterial diseases of potato plants (black leg) and tubers (black leg, bacterial rot, and bacterial ring rot) were studied in the same way.

Visually assessed viroses included rugose mosaic, leaf drop streak, and leaf rolling mosaic. During the period of budding and the onset of blooming serological tests for viruses X, S and M were carried out using 30% of potato plants.

Delineation of Response Range Surfaces for the Studied Potato Parameters

The complex diagnostics of strain-phytopathogen interactions was carried out using the "rapid ecological crop trial (RECT)" approach (Stefanova, 2002). The key aspects of this approach are shown in the diagram presented in Fig. 1. The article describes data about the crop yields of cultivars and the results of phytosanitary monitoring.

Statistical software packages were used to describe the response ranges of potato strain samples with regard to their productivity and resistance to phytopathogens as influenced by the two environmental parameters, HTC and NPK.

After tabulated database integrating all variants of experiments and all parameters studied had been built, the effects of the factors under study on parameters of interest were evaluated. Thereafter, surface plots were built in XYZ coordinates. The potato strains were ranked by their ranges of responses to the environmental parameters HTC and NPK. In a group of five strains, rank 1 was assigned to the strain with the greatest response range and rank 5 to the strain with the smallest response range. Thereafter the mean value of ranks for a group of characteristics related to mycoses, bacterioses, viroses and the susceptibility of leafy tops and tubers to diseases was calculated for each strain. The article presents data treated according to rapid ecological crop trial (RECT) approach and related to the first five of the aforementioned potato strains.

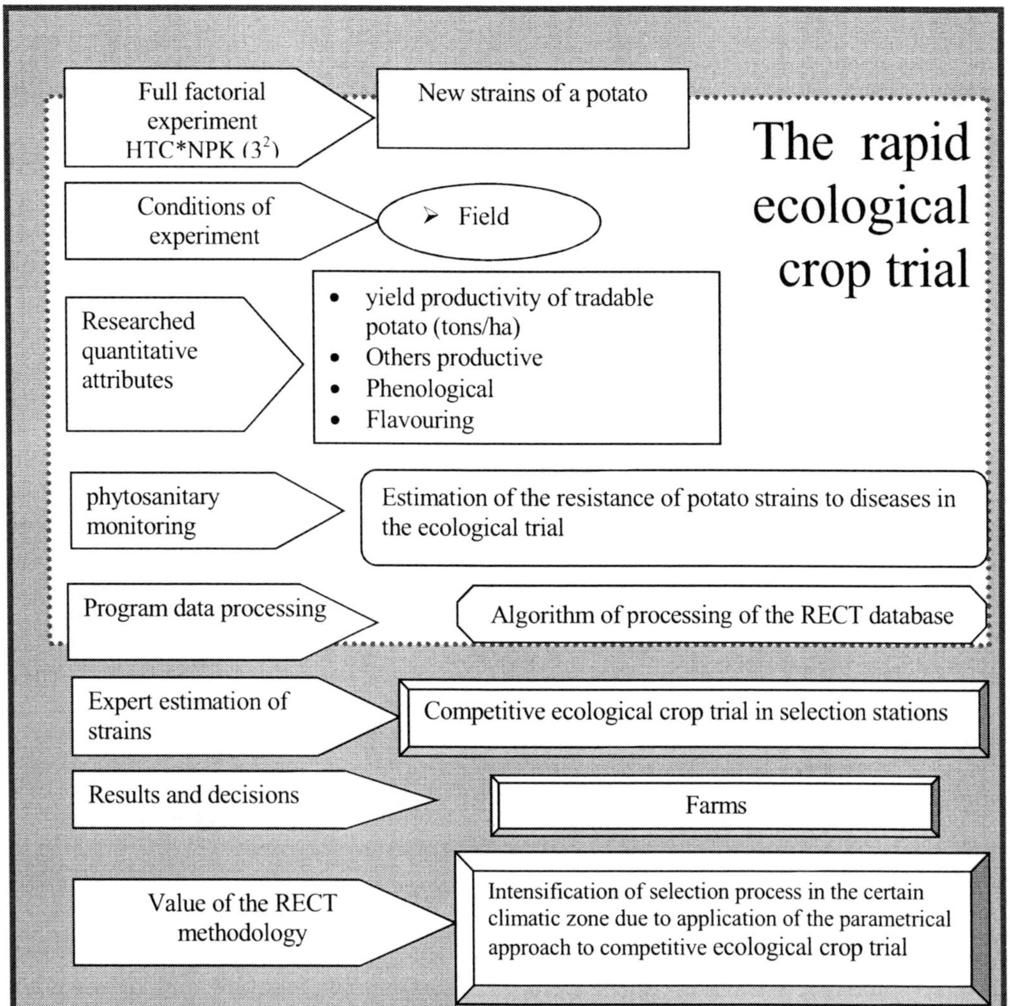

Fig. 1. Flow chart of RECT use exemplified by row crops.

RESULTS AND DISCUSSION

Computer Database for Express Cultivar Trails

The success of row crops trials depends on adequate experimental design and the conformance of trial conditions to production conditions including phytopathogen burden and the hydrothermal characteristics.

Modern farms use computer bases of data and knowledge and mathematical models for prediction of productivity and for planning of planting process (Annevelink et al., 1993). We have developed a database for ecological trials of potato strains taking account of not only optimal planting zones but also of non-optimal planting regimens. Trial data are reduced to mean values and mean errors using statistical software packages and then tabulated.

At present, the models believed to be the most relevant are parametric models taking account of limitations in humidity and nitrogen nutrition of plants (Kabat et al., 1995). Therefore we suggested a solution system that takes account of such working factors as HTC and NPK and, also, of the precedent crop and of planting zone (region).

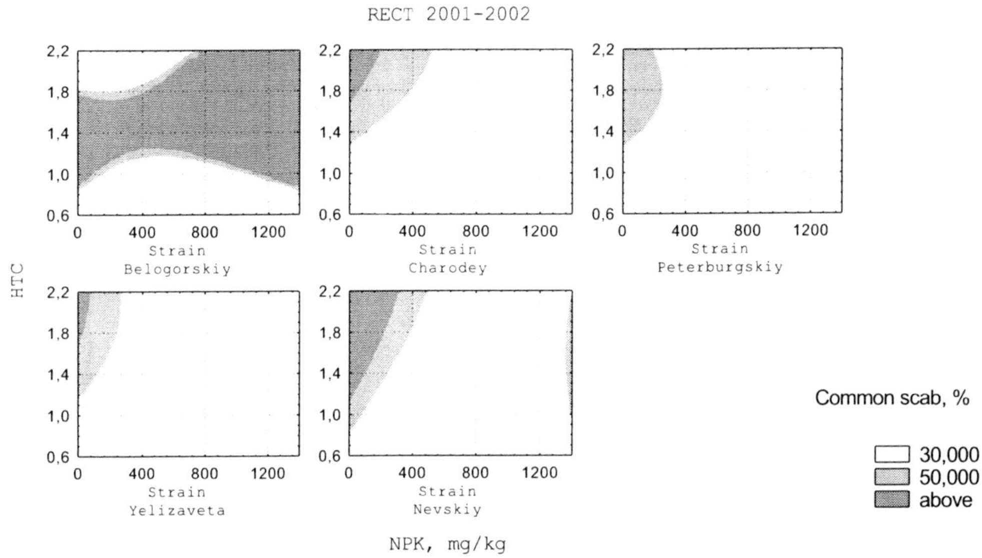

Fig. 2. Response ranges for susceptibility of five potato strains to common scab, %.

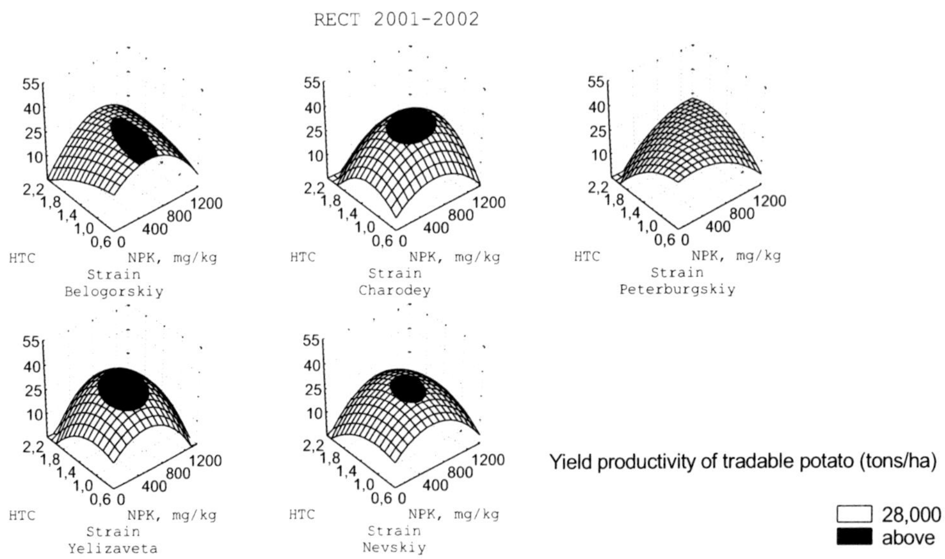

Fig. 3. Results of RECT for five potato strains.

At first, the complete statistical treatment of raw data is carried out, which includes checking of the homogeneity and continuity of distributions and statistical assessment of the significance of the effects produced by different factors.

The chosen regression model for potato responses to combinations of agrometeorological conditions was the standard square polynomial regression and least square approximation of data.

The ranges of productivity responses of strains under study and of their resistance to diseases were used as criteria for expert data evaluation and generalization and prognostication (Fig. 2 and 3).

The Resistance of Potato Strains to Diseases in the Ecological Trial

According to the results of multifactorial analysis of variance, the leading factor in the development of potato diseases was trial site. The data obtained in phytosanitary monitoring of five potato strains and five trial sites are presented in Table 2.

The greatest rate of lesioning by common scab has been found at Site A. This trial site is the northern extreme of potato agriculture in Russia (Kostiuk et al., 2001). However, the finding is most likely caused not only by inadequate weather conditions and climatic characteristics but also by the fact that potato was cultivated at the site prior to the trial. Yelizaveta and Peterburgskiy potato strains proved to be the most scab-resistant (Fig.2). Much attention was given to assessment of the resistance of the tested strain samples to viral diseases including late blight, dry scab, leaf spot, early blight, and black dot, which greatly reduce tuber yields.

At other sites, all strains showed very good growth and shrub development, high germinating capacity, the absence of fungal diseases, and fairly good tuber yield. The trial has shown that the tested potato strains will be in demand in regions that feature drastic weather changes promoting late blight development.

To determine the specificity of their complex resistance, the potato strains were ranked according to their ranges of responses to phytopathogens (Table 3). In this way the Peterburgskiy strain was singled out as the most tolerant to mycoses and bacterioses and the Charodey strain, to viroses, Charodey featuring a complex resistance to infections of both leafy tops and tubers. Peterburgskiy showed the greatest range of responses with regard to complex resistance to tuber diseases. The resistance of Charodey and Peterburgskiy strains to tuber diseases makes them the most promising for seed-farming because tuber diseases are mainly responsible for compromised validity of selection materials (Struik and Wiersma, 1999).

The yield was higher when the susceptibility to diseases was lower. Black scab, late blight at its onset, and common scab decreased tuber yield and marketability and the number and net mass of marketable tubers, correlation coefficients being significant and ranging from –0.4 to –0.6. Bacterioses affected the quality of tubers and their yield. Viruses Y, S, and M affected tuber mass per plant. Virus X combined with virus S adversely affected all characteristics of productivity considered herein.

As to potato taste quality, bacterioses compromise tuber flavor and color, viroses adversely influence tuber odor and cooking properties, and mycoses (leaf spot, early blight, black dot, dry scab, black scab, and skin necrosis) spoil all degustation characteristics.

Potato potential for high tuber yields within a broad range of ecological conditions is very important for producers. So studies of the efficacy of selection depending on environmental conditions are of a great importance for selection process.

Table 2. Characteristics of the resistance of potato plants to phytopathogens in ecological crop trial of 5 potato strains in the Northwest of Russia.

		Belogorskiy					Charodey					Peterburgskiy					Yelizaveta					Nevskiy				
	strains Diseases	A	B	C	D	E	A	B	C	D	E	A	B	C	D	E	A	B	C	D	E	A	B	C	D	E
mycoses	Rhizoctonia disease in 10 days after flowering, a point	7	9	9	6	9	7	9	9	6	9	8	9	9	7	9	7	8	9	7	9	7	8	9	7	8
	Macrosporium disease in 10 days after flowering, a point	9	9	9	7	9	9	9	9	8	9	9	8	9	7	9	7	9	9	7	9	9	9	9	7	9
	Alternaria disease in 10 days after flowering, a point	9	9	9	9	9	9	9	9	8	9	9	5	9	8	9	7	5	9	8	9	9	5	9	8	9
	Anthracnose in 10 days after flowering, a point	9	9	9	8	9	9	9	9	9	9	9	5	9	9	9	9	5	9	9	9	9	5	9	9	9
	Late blight - The moment of occurrence, point	7	9	9	8	9	7	9	9	9	9	9	9	9	8	9	7	9	9	8	9	9	9	9	8	9
	Late blight - 10 days from the moment of occurrence, a point	7	9	7	7	9	7	9	8	8	9	9	9	8	7	9	7	9	7	8	8	9	9	8	7	9
	Late blight - 20 days from the moment of occurrence, a point	7	9	9	7	9	7	9	9	8	9	9	9	8	7	9	7	9	5	7	8	9	9	7	7	8
	Common scab on tubers in middle of winter, defeat in %	59	0	2	2	0	42	0	2	0	0	27	0	0	0	0	25	0	9	0	0	39	0	31	1	6
	Silver scurf on tubers in middle of winter, defeat in %	0	0	0	0	0	0	0	4	0	0	0	0	0	1	0	0	0	0	0	0	0	0	0	0	0
	Rhizoctonia disease on tubers in middle of winter, defeat in %	0	0	10	4	0	0	0	0	1	0	0	0	3	3	0	0	0	2	2	0	2	0	2	2	0
	Phomosis on tubers in middle of winter, defeat in %	1	0	0	0	7	0	0	0	0	1	0	0	0	1	1	0	0	0	0	7	1	0	0	0	5
	Late blight on tubers in middle of winter, defeat in %	2	1	2	2	3	0	0	0	0	0	0	0	0	0	0	0	0	0	1	0	0	0	0	0	0

Table 2. Continued

		Belogorskiy					Charodey					Peterburgskiy					Yelizaveta					Nevskiy				
	strains Diseases	A	B	C	D	E	A	B	C	D	E	A	B	C	D	E	A	B	C	D	E	A	B	C	D	E
bacterioses	Black leg disease in 10 days after flowering, a point	7	7	9	7	7	9	9	5	5	5	9	9	9	9	9	9	9	5	5	9	9	9	9	9	9
	Black leg on tubers in middle of winter, defeat in %	0	0	0	1	1	0	0	0	0	0	0	0	0	0	0	0	0	0	0	0	0	0	0	0	0
	Bacterial ring rot on tubers in middle of winter, defeat in %	0	0	0	1	1	0	0	0	0	0	0	0	0	0	0	0	0	0	0	0	0	0	0	0	0
	Bacterial rot on tubers in middle of winter, defeat in %	1	0	0	0	0	0	0	0	0	0	0	0	0	0	0	0	0	0	0	0	0	0	0	0	0
viroses	Rugose mosaic, %	0	0	0	0	0	0	0	0	0	0	0	0	0	0	0	0	0	0	0	0	1	0	1	3	1
	Leaf drop streak, %	0	0	0	0	0	0	0	0	0	0	0	0	0	0	0	0	0	0	0	3	0	0	0	0	0
	Mosaic ordinary, %	0	0	0	0	0	0	0	0	0	0	0	0	0	0	0	1	0	0	0	2	4	1	2	9	4
	Leaf rolling mosaic, %	0	0	0	0	0	0	0	0	0	0	0	0	0	0	0	2	3	1	2	3	0	0	0	0	0

Table 3. Ranks of the complex resistance of potato strains to mycoses, bacterioses and viroses determined by RECT in 2001-2002 and mean yield productivity of tradable potato (tons/ha) at the control level of fertilizes.

Strain	Ranks of the complex resistance					yield productivity of tradable potato (tons/ha)		
	mycosis	bacteriosis	virosis	leafy tops	tubers	2001	2002	2003
Belogorskiy	3	3	2	1	5	26,4	24,0	24,5
Charodey	2	2	1	2	2	24,8	25,6	29,7
Peterburgskiy	1	1	3	3	1	29,0	24,4	
Yelizaveta	5	4	4	5	3	27,3	27,1	28,8
Nevskiy	4	5	5	4	4	29,1	24,5	26,2

Our studies conducted in 2003 under conditions of early epiphytic development of late blight and spring favorable for black scab yielded potato strains that proved to be relatively tolerant to the disease that are prevalent in the North and Northwest of Russia. On the whole, the most favorable conditions 20 days before harvesting for yielding good crops for strains Nevskiy, Yelizaveta, and Charodey (near 12.4 tons/ha) were observed at Sites D and E. Similar results were obtained during the main harvesting period. The strains Yelizaveta proved to be the most productive at Site D (40.2 tons/ha), Nevskiy, Yelizaveta, and Charodey at site A (37.2 – 39.8 tons/ha), and Charodey at Site E (37.1 tons/ha).

An important characteristic of any present time potato sort is its ecological plasticity implying the responsiveness of its genotype to environmental changes and the stability of the main parameters of its productivity, which include tuber yield in the first place (Sapega, 2002). The strains under study displayed different tuber yield response ranges (Fig. 3). The plots presented herein suggest that Yelizaveta strains are suitable for a wider zoning. Charodey strain proved to be vigorous yet having a narrow response range, and thus it may be introduced only in separate zones satisfying specific requirements for soil quality and climate.

Conclusions

Complex diagnostics of susceptibility to diseases in ecological trials of cultivars basing on the RECT approach will allow determining of cultivars and strains that are tolerant to prevalent natural phytopathogens. The method suggested herein is based on new generation technologies for expert evaluation and provides for the following:

(1) Saving of time elapsing from the development of a novel cultivar to its assigning to specific zones.
(2) Saving of resources needed for preliminary trials.
(3) Obtaining of more reliable prognoses of productivity under known conditions (climate and land treatment).

This approach may be useful for speeding up of selection at the stage of comparative trials of candidate strains for determination of the most adapted of them and for prognostication of their further use.

The studies suggest that the strain Yelizaveta may be referred to the category of ecologically plastic genotypes, Charodey may be categorized both as an complex phytopathogen resistance strain, and as intensive type strain showing strong positive responses to improved environmental conditions yet reduced productivity under unfavorable conditions.

Acknowledgments

The authors are thankful to Nina A. Neliubina, Olga V. Balychova, and Nikolai I. Tsarkov for kindly provided data of regional trials and for valuable comments to the method of rapid ecological crop trial.

REFERENCES

[1] T. Lorkh: *Ecological Plasticity of Potato Strains*. Kolos, Moscow, 1968. 32 p. (In Russian).

[2] P. Losev: *Practical Manual of Agrometeorological Securing of Plant Cultivation*. Gidrometeoizdat, Saint-Petersburg, 1994. 244 p. (In Russian).

[3] N. A. Stefanova: Fungi-resistance, potential and real productivity in the rapid ecological crop trial. *Abstracts of 8th New Phytologist Symposium*. HU Press, Helsinki, 2002. P.046.

[4] K. M. Swiezynski: Some interesting aspects of potato breeding in Europe. *Acta Hort.*, 213, 51 (1987).

[5] Schüber, E. Schiessendoppler: Some interesting aspects of recent and expected developments of diseases and pests and their control in Europe. *Acta Hort.*, 213,169 (1987).

[6] P. C. Struik, S. G. Wiersema: *Seed potato technology*. WAP, Wageningen, 1999. 384p.

[7] P. Kabat et al. (Eds.): *Modelling and parametrization of the soil-plant-atmosphere system: a comparison of potato growth models*. WAG, Wageningen, 1995. 513p.

[8] E. Annevelink et al. (Eds.): *Farm planning, labour and labour conditions, computers in agricultural management*. WAG, Wageningen, 1993. 420p.

[9] H. P. Beukema, D. E. Van Der Zaag: *Introduction to Potato Production*. PUDOC, Netherlands, 1990, 207 p.

[10] S. Kiru: Status of the VIR potato genebank. *Report of a working group on potato*. IPGRI, Rome, 2001, P.57.

[11] V.A. Sapega: Efficiency and ecological plasticity of strains of a potato in Northern Ural. *Agrarian science*, 11, 18 (2002). (In Russian).

[12] V. I. Kostiuk et al.: Influence of terms of landing on size and quality of a crop of a potato on Kola peninsula. *Agrarian science*, 1, 16 (2001). (In Russian).

In: Biotechnology in Agriculture and the Food Industry
Editor: G. E. Zaikov, pp. 63-69
ISBN 1-59454-119-1

Chapter 9

DNA GENOTYPING OF THE POTATO AND ITS WILD RELATIVES

M. P. Beketova[1, 2], V. A. Biryukova[1, 2], N. S. Velishaeva[1], V. S. Zaitsev[1], A. A. Pankin[1], E. E. Khavkin[1], L. M. Khromova[2], O. V. Vanyusheva[1] and I. A. Shilov[1]

[1] Institute of Agricultural Biotechnology, 42 Timiryazevskaya ul., Moscow 127550 Russia; fax: 7(095) 977-0947; e-mail address: emil@iab.ac.ru

[2] Institute of Potato Husbandry, Korenevo, Moscow region, 140052 Russia

ABSTRACT

Both moderately repetitive R173 sequences and simple sequence repeats adequately discern *Solanum* species and subspecies and potato varieties, hybrids, lines, and somaclones. In the case of derived potato cultivars and mericlones, the balance of advantage lies with the SSR analysis. Together two technologies provide over 100 DNA descriptors sufficient for reliable identification of any *Solanum* genotypes.

Key words: *Solanum*; DNA genotyping; cultivar identification; somaclonal variation; moderately repetitive sequences; simple sequence repeats; microsatellites

The technologies of DNA genotyping provide a reliable instrument for characterizing and classifying plant species, subspecies, varieties, inbred lines, and clones and evaluating their genetic and allelic diversity. DNA profiles are essential for maintaining the existing germplasm collections and adding new forms with economically important genes or alleles. By estimating the genetic similarity or genetic distance between genotypes, one can establish cultivar parentages. These technologies can considerably advance the breeding process due to efficient capturing the prospective genotypes and genes for introgression and by rapid monitoring the transfer of target genes in segregating progenies of wide crosses. DNA profiling technologies would protect breeders' rights and promote the cultivar identification and registration, the control of seed purity and uniformity, and the assurance of food safety, including the certification of genetically modified plants and their products [1].

When establishing new technologies of DNA genotyping, one should consider the comparative advantages and disadvantages of DNA markers employed as descriptors for plant fingerprinting, the comparative technical specifications of analytical methods, and the cost-efficiency of large-scale genotyping programs. Having these criteria in mind, we compare a comparatively new technology of plant genotyping on the basis of moderately repetitive DNA sequences (retrotransposons) [2] with the already established high throughput technology of microsatellite (simple sequence repeat, SSR) analysis [1].

In their study of the family R173 dispersed repeats in rye and wheat, Rogowsky et al. [4, 5] designed PCR primers for synthesizing the dominant markers suitable for genome mapping and cytogenetic analysis. In each pair of these primers, one recognizes the repetitive R173 sequence, and another, the flanking region of the genome. The R173 repeats are 3-5 Kbp in length, with the copy number about $15\ 10^3$ per diploid rye genome; besides, the study of barley genome [6] revealed that microsatellite sequences within the R173 repeats can add up to their polymorphism.

Previously we demonstrated that the primers and PCR protocol developed by Rogowsky et al. [4, 5] could be used to discern and identify potato varieties [6, 7]. The primers for microsatellite analysis have been chosen from the published database [8-10], and the protocols of amplification were verified in our laboratory. Genomic DNA was isolated from young leaves by a modified CTAB method. The R173 amplicons were separated by electrophoresis in 1.5% agarose gel containing 10 mg/ml ethidium bromide in 1x TAE buffer at 22 V/cm. SSRs were separated in 4.5% polyacrylamide gel using an ALFexpress II analyzer (Amersham BioSciences, United Kingdom) and an ALFwin Fragment Analyzer 1.00 software. Our study comprised over 70 cultivated and wild *Solanum* forms, including the somaclonal lines obtained from the callus culture and the isolated meristem culture (mericlones). All plant material was obtained from the working collections of the Institute of Potato Husbandry, Korenevo, Moscow region.

The profiles of the family R173 repeats and SSRs varied conspicuously among *Solanum* species and subspecies; especially high was the genetic diversity within the forms of *S. tuberosum* (Figs. 1-3). In several cases, the moderate repeat technology failed to discern closely related (derived) potato cultivars; in these cases, the SSR technology was more successful.

By comparing wild *Solanum* genotypes with potato varieties related to these particular species by known pedigrees, we could discern the common genome fragments and the characteristic genome fragments found only in the wild species (Figs. 2 and 3). Of particular interest are the DNA fragments apparently transferred from wild *Solanum* species into potato varieties. First, such descriptors provide DNA markers of introgressed genetic material essential for rapid assessment of the segregating progenies of wide crosses. Second, some of these moderate repeats may comprise the fragments of the introgressed target genes. To illustrate, when we cloned the R173 repeats from *S. chacoense* and *S. kurzianum*, they comprised similar 529-bp stretches 85-89% identical to a non-transcribed region of the *R1* gene of resistance to late blight cloned from *S. demissum* (the GenBank accession AF447489 [11].

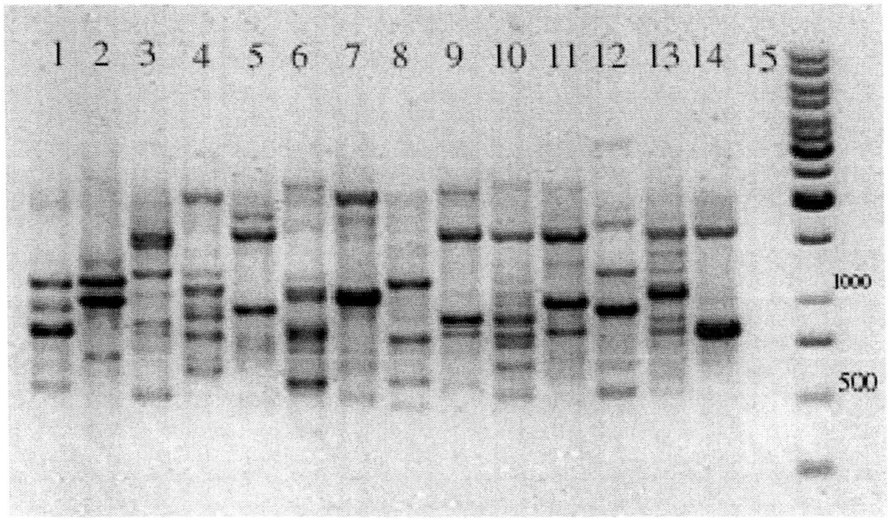

Fig. 1. Profiles of the R173 moderate repetitive sequences in Solanum species. Amplification with the primer PawS6. Here and below, in parentheses are CIP and VIR catalog numbers.1 - S. nigrum (L.) Em. Mill., 2 – S. lycopersicoides Dunal (Ser. Juglandifolia), 3 – S. boliviense Dunal (PI 310928) (Ser. Megistacroloba), 4 – S. megistacrolobum Bitter. (Ser. Megistacroloba) (PI 275148), 5 - S. chacoense Bitter. (PI 762268) (Ser. Yungasensa), 6 – S. demissum Lindl. (PI 275208) (Ser. Demissa), 7 – S. tuberosum L. ssp. andigena (Juz. et Buk.) Hawkes (PI 603632) (Ser. Tuberosa), 8 – S. stoloniferum Schltdl.et Bouchet.(PI 255532) (Ser. Longipedicellata), 9 - S. microdontum Bitter. (PI 762341) (Ser. Tuberosa), 10 – S. marinasense Vargas (PI 310946) (Ser. Tuberosa), 11 – S. verrucosum Schltdl. (K-21695) (ser. Tuberosa), 12 – S. immite Dunal (PI 365330) (Ser. Tuberosa), 13 - S. sucrense Hawkes (PI 473367) (Ser. Tuberosa), 14 – S. kurtzianum Bitter. et Wittm. (PI 472923) (Ser. Tuberosa), 15 – control without DNA, 16 – molecular weight marker 1kb DNA Ladder.

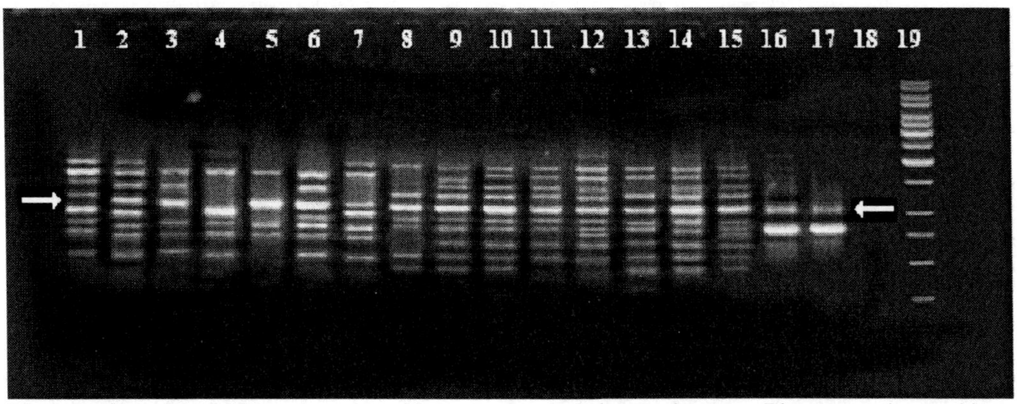

Fig. 2. Profiles of the R173 moderate repetitive sequences in Solanum chacoense Bitter. accessions and in potato varieties related to S. chacoense. Amplification with the primer PawS6. 1 - S. chacoense (PI 189219); 2 - S. chacoense (PI 133713); 3 - S. chacoense (PI 472828); 4 - S. chacoense (PI 472832); 5- S. chacoense (PI 472810); 6 - S. chacoense (PI 762268); 7 - S. chacoense (PI 265576); 8-17 – S. tuberosum: 8 - Lugovskoi 20; 9 - Lugovskoi 700; 10 - Zhukovskii VIR; 11 – Zhukovskii rannii 25; 12 - Belosnezhka; 13 – Atlantic; 14 – Il'inskii 30; 15 - Il'inskii 23; 16 - Nikulinskii 8; 17 - Nikulinskii 1; 18 – control without DNA; 19 - molecular weight marker 1kb DNA Ladder. Arrows indicate the DNA fragment presumably transferred from S. chacoense to S. tuberosum.

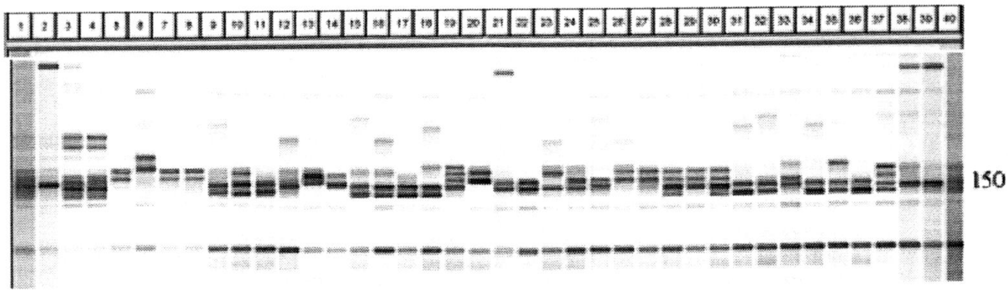

Fig. 3. Microsatellite profiles of Solanum species and potato varieties. Primers POT 83 and POT 84 [10]. Solanum species: 3, 4 - S. nigrum, 5 - S. lycopersicoides, 6 - S. kurtzianum, 7, 8 - S. stoloniferum, 14, 15 - S. chacoense, 22 - S. demissum. Potato varieties related to S. stoloniferum: 9 – Bryanskii rannii, 10 - Skoroplodnyi, 11- Effekt, 12 - Golubizna, 13- Resurs. Potato varieties related to S. chacoense: 16 - Belosnezhka, 17 - Zhukovskii rannii, 18 - Atlantic, 19 – Il'inskii, 20 - Lugovskoi. Other potato varieties: 23 - Udacha, 24- Sinetsvet, 25- Luk'yanovskii, 26 - Peterburgskii, 27 - Nevskii, 28- Kameras, 29 – Smena-20, 30 -Desiree, 31- Romano, 32- Katahdin, 33 -Russet Burbank, 34 - Kennebec, 35 - Wauseon, 36 - Lady Rosetta, 37 - Maris Piper, 38 – Saturna. 2, 21 and 39 - molecular weight marker, bp.

The analysis of the R173 repeats discerned variation in at least some of potato somaclonal lines selected in the callus tissue culture (Fig. 4a). This evidence is of particular interest as regards the presumable role of retroelements in somaclonal variation [12]. However, the moderate repeat technology was not as helpful in the case of mericlone lines produced in the isolated meristem culture. In this case, the balance of advantage lies with the SSR analysis: at least in two cases, e.g., with var. Golubizna and Lugovskoi, mericlones, were distinct from the initial forms (Fig. 4b).

To prepare DNA profiles of *Solanum* genotypes, each well-resolved DNA fragment of known molecular weight was scored as a particular descriptor. Matrices comprising the data about the presence or absence of these descriptors were prepared for all genotypes under study and all primer combinations. A computer program TREECON was used to calculate the genetic distances between genotypes after Nei и Li [13] and to produce, by the UPGMA method, the dendrograms, which were compared to the phylogenetic relations of the genotypes under comparison and/or their pedigrees.

Evidence from the phylogenetic tree based on the R173 descriptors (Fig. 5) is in good agreement with the known pedigrees of potato varieties. In particular, we separated most potato varieties derived from either *S. chacoense* or *S. stoloniferum*. We could also discern between potato varieties bred in Russia and abroad, e.g., Zhukovskii rannii, Golubizna, Effekt, Il'inskii, and Bryanskii rannii vs. Russet Burbank, Wauseon, Kennebec, Katahdin, Romano, and Desiree. The association of var. Il'inskii and Udacha with var. Desiree is substantiated by their pedigrees. The characteristic position of var. Atlantic can probably stem from the fact that this cultivar comprises the genetic material from *S. chacoense*. The most characteristic feature of the phylogenetic tree based on the SSR descriptors (not shown) was sharp separation of wild and cultivated *Solanum* forms, except *S. demissum* related to most potato varieties.

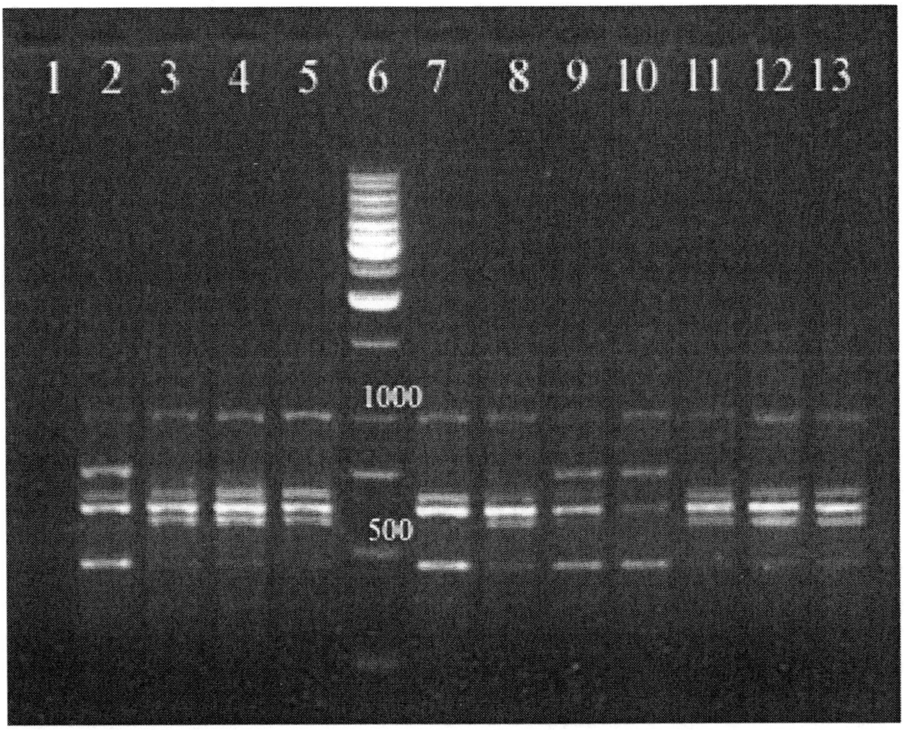

(a)

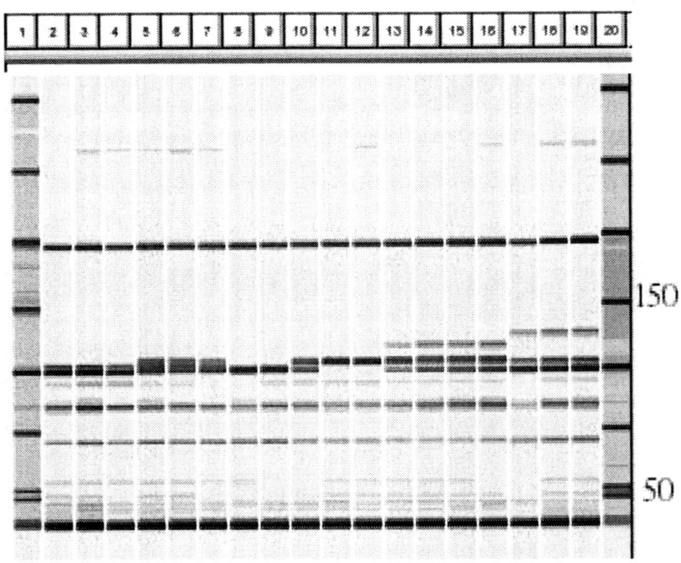

(b)

Fig. 4. Somaclonal variation in potato. (a) Profiles of the R173 moderate repetitive sequences in the initial potato forms (IF) and somaclones obtained from the isolated callus culture. Primers Paw5 + Paw16. 1 – Control without DNA; 2 - 9к-29 IF; 3-5– somaclonal lines 97, 101 and 104; 6 – molecular weight marker 1 kb GeneRuler DNA Ladder; 7 – Smena-20 IF; 8-13 – somaclonal lines 1*, 1, 2, 4, 8 and 7. (b) Microsatellite profiles in the initial potato forms (IF) and somaclones obtained from the isolated meristem culture. Primers POT 83 and POT 84 [10]. 1, 20 – molecular weight marker, bp; 2 – Zhukovskii rannii IF; 3, 4 – mericlonal lines Zhukovskii rannii VIR and 25; 5 – Il'inskii IF; 6, 7 - mericlonal lines Il'inskii 30 and 23; 8 – Golubizna IF; 9, 10 – mericlonal lines Golubizna R-25 and 80; 11 – Lugovskoi IF; 12, 13 – mericlonal lines Lugovskoi 700 and 20; 14 – Udacha IF; 15, 16 – mericlonal lines Udacha 11 and 13; 17 – Nikulinskii IF; 18, 19 – mericlonal lines Nikulinskii 8 and 1.

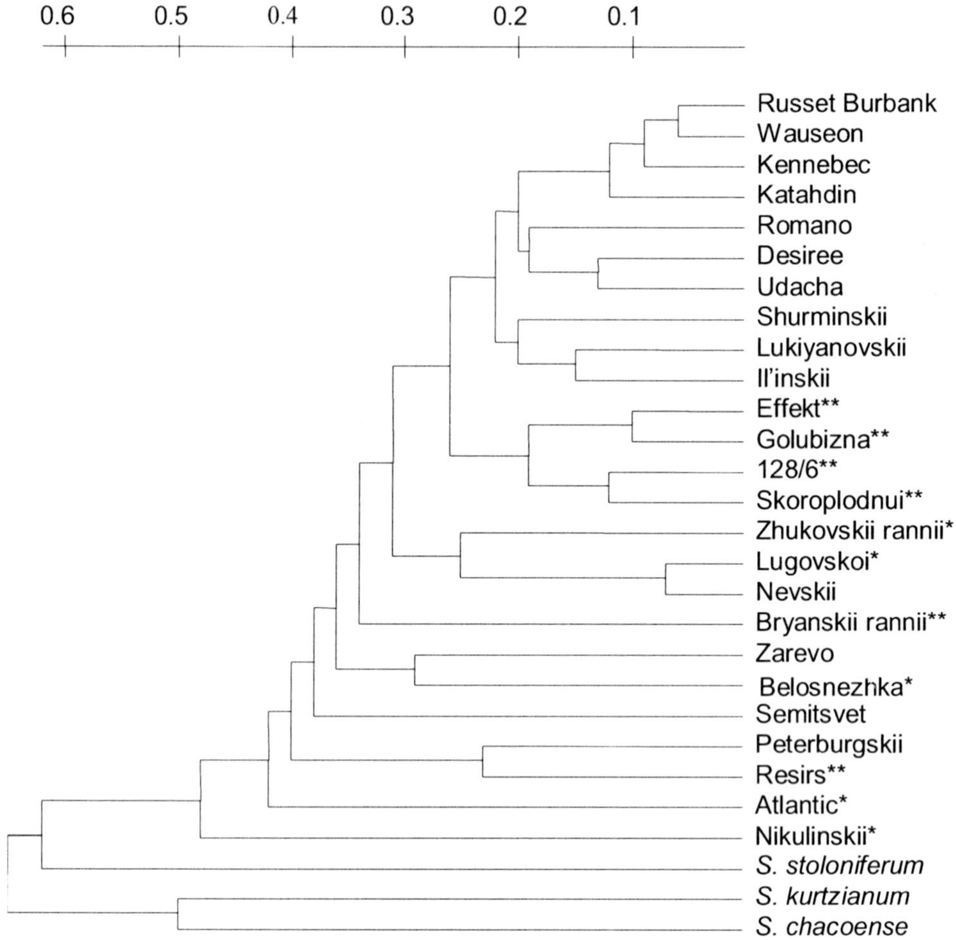

Fig. 5. The phylogenetic tree of potato genotypes built on the basis of the R173 moderate repetitive sequences. On the abscissa axis, genetic distances after Nei and Li [13]. *Potato varieties related to S. chacoense; ** potato varieties related to S. stoloniferum.

The R173 technology is simpler and less costly than the SSR technology. When combined, two technologies provide over 100 DNA descriptors sufficient for reliable identification of any cultivated and wild *Solanum* genotypes.

References

[1] R.J. Henry (Ed.). *Plant genotyping: The DNA Fingerprinting of Plants*. Wallingford, UK, 2001.

[2] A. Kumar, H. Hirochika. Applications of Retrotransposons as Genetic Tools in Plant Biology. *Trends Plant Sc.,* 6 (3), 127 (2001).

[3] P.M. Rogowsky, K.W. Shepherd, P. Langridge. Polymerase Chain Reaction Based Mapping of Rye Involving Repeated DNA sequences. *Genome*, 35 (4), 621 (1992).

[4] P. Rogowsky, J.-Y. Liu, S. Manning, P. Langridge. Structural Heterogeneity in the R173 Family of Rye-Specific Repetitive DNA Sequences. *Plant Mol. Biol.,* 20 (1), 95 (1992).

[5] L. Ramsay, M. Macaulay, L. Cardle et al. Intimate Association of Microsatellite Repeats with Retrotransposons and Other Dispersed Repetitive Elements in Barley. *Plant J.,* 17 (4), 415 (1999).

[6] V.S. Zaitsev, E.E. Khavkin. Identification of Plant Genotypes by the PCR Analysis of R173 Dispersed Repetitive Sequences. *Doklady RASKhN*, no. 2, 3 (2001).

[7] V.A. Biryukova, V.S. Zaitsev, E.E. Khavkin et al. DNA Markers in Potato Breeding. *Dostizheniya nauki i techniki APK*, no. 10, 38 (2003).

[8] L.M. Kawchuk, D.R. Lynch, J. Thomas et al. Characterization of *Solanum tuberosum* Simple Sequence Repeats and Application to Potato Cultivar Identification. *Amer. Potato J.*, 73 (4), 325 (1996).

[9] D. Milbourne, R.C. Meyer, A.J. Collins et al. Isolation, Characterisation and Mapping of Simple Sequence Repeat Loci in Potato. *Mol. Gen. Genet.,* 259 (3), 233 (1998).

[10] V. Ashkenazi, E. Chani, U. Lavi et al. Development of Microsatellite Markers in Potato and Their Use in Phylogenetic and Fingerprinting Analyses. *Genome*, 44 (1), 50 (2001).

[11] A. Ballvora, M.R. Ercolano, J. Weiss et al. The R1 Gene for Potato Resistance to Late Blight (*Phytophthora infestans*) Belongs to the Leucine Zipper/NBS/LRR Class of Plant Resistance Genes. *Plant J.,* 30 (3), 361 (2002).

[12] S.E. Kubis, A.M.M.F. Castilho, A.V. Vershinin, J.S. Heslop-Harrison. Retroelements, Transposons and Methylation Status in the Genome of Oil Palm (*Elaeis guineensis*) and the Relationship to Somaclonal Variation. *Plant Mol. Biol.*, 52 (1), 69 (2003).

[13] M. Nei, W.H. Li. Mathematical Model for Studying Genetic Variation in Terms of Restriction Endonucleases. *Proc. Natl. Acad. Sci. USA*, 76 (10), 5269 (1979).

In: Biotechnology in Agriculture and the Food Industry ISBN 1-59454-119-1
Editor: G. E. Zaikov, pp. 71-76

Chapter 10

DNA Markers of Developmental Genes in Brassicaceae Crop Plants

V. V. Martynov and E. E. Khavkin
Institute of Agricultural Biotechnology, Russian Academy of Agricultural Sciences,
Timiryazevskaya ul., 42, Moscow, 127550 Russia;
fax: 7 (095) 977-0947; e-mail: emil@iab.ac.ru

Abstract

Fragments of the candidate genes that control morphogenesis and the rate of development were cloned and characterized from the *Brassica, Camelina*, and *Solanum* crop plants. The probes based on these genes cross-hybridized to *Brassica* and *Solanum* DNA.

Key words: plant development; meristem; flowering; DNA markers; *Brassica; Solanum;* receptor-like kinase genes; MADS-box genes

Many genes governing plant morphogenesis and the rate of development have been recently characterized in arabidopsis, rice, maize, tomato and other species. However, the genes that determine the architecture of the shoot, photosynthetic canopy, inflorescence, and flower, the rate of development, and the yield structure have not been identified in many economic plants. Working on the idea that these candidate genes are the structural homologs of the already known genes (prototype genes) with similar physiological functions in the systematically related species can considerably help their identification.

The family Brassicaceae comprises many food and fodder species of diverse genome constitutions and growth habits. Many genes that determine this variation were described and mapped only by their phenotypic manifestations. The current research on arabidopsis genes makes the candidate-gene concept especially beneficial for developing DNA markers of the developmental genes/traits in the Brassicaceae crop plants [1, 2].

Our approach to developing these markers is based on the comparative study of plant genomes rather than cDNAs for the particular genes: numerous mutations in introns were reported to affect the plant morphology or the rate of development.

Following PCR amplification of genomic DNA with the primers recognizing the conserved stretches in the arabidopsis prototype genes, we cloned gene fragments thus obtained and identified them by their primary structure (Table 1).

Table 1. Cloned fragments of the homologs of arabidopsis developmental genes

Prototype gene	Plant source	Fragment size, bp	% homology to the prototype gene	NCBI GenBank accession
CLAVATA1	*Camelina sativa*	509	91	AF467952
Same	*Brassica napus*	509	93	AY130760
Same	*Brassica rapa*	509	91	AY130759
Same	*Brassica napus*	3008	88	AY283519
ERECTA1	*Camelina sativa*	549	94	AY069973
BRASSINOSTEROID INSENSITIVE1	*Camelina sativa*	567	91	AF489487
CONSTANS	*Brassica rapa*	705	60	AY379531
Same	Same	705	61	AY379532
FLOWERING LOCUS C	*Brassica juncea*	1588	84	AY266265
Same	Same	1688	84	AY268931
SHORT VEGETATIVE PHASE	*Camelina sativa*	521	92	AY177710
SUPPRESSOR OF OVEREXPRESSION OF CONSTANS1	*Solanum tuberosum*	344	61	-
Same	Same	344	62	-

The hybridization probes produced from the structural gene homologs were used to assess by RFLP analysis whether the putative candidate genes associate with the phenotypic manifestations of the developmental traits. By analyzing the restriction site polymorphism, one can establish such association even with the hybridization probes recognizing the whole group of structurally related genes.

Three genes, *CLAVATA1* (*CLV1*), *ERECTA1* (*ER1*), and *BRASSINOSTEROID INSENSITIVE1* (*BRI1*) (Fig. 1), belong to the *Xa21* subfamily of the receptor-like kinase genes [3]. The product of these genes comprises three domains (Fig. 2): an extracellular LRR domain containing leucine-rich repeats, a transmembrane domain, and an intracellular kinase domain. The homologs of *BRI1, CLV1*, and *ER1* isolated from the Brassicaceae crop plants were very similar to the prototype genes (see Table 1).

Among these genes, particularly interesting is *CLV1*, a component of the *CLAVATA-WUSCHEL* regulatory contour [4]. This contour maintains the balance between the stem cells in the central zone of stem apical meristem (SAM) and the cells at the SAM periphery already primed for determination, identification, and differentiation of leaf and stem tissues [5].

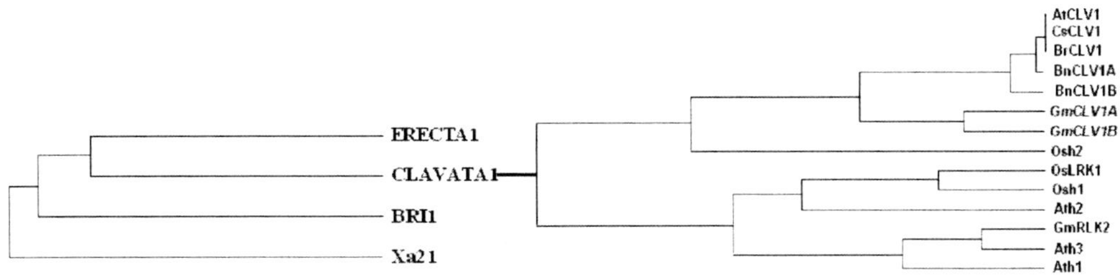

Fig. 1. The phylogenetic tree of the Xa21 subfamily of receptor-like kinases. On the right, the comparison of the arabidopsis CLAVATA1 prototype and its homologs within and beyond the Brassicaceae family. AtCLV1 - the arabidopsis CLAVATA1 (GenBank accession no. Q9SYQ8), Ath1, Ath2, and Ath3 - arabidopsis receptor protein kinases (CAB66905, CAB79027, and AAM20665), GmCLV1A and GmCLV1B - CLAVATA1 homologs (AAF59905 and AAF59906) from Glycine max, GmRLK2 - receptor kinase (AAF91323) from G. max, Osh1, Osh2, and OsLRK1 - receptor kinases (AAM27467, AAL79717, and AAF73373) from Oryza sativa, BnCLV1A and BnCLV1B - CLAVATA1 homologs (AAN05448 and AAP49010) from Brassica napus, BrCLV1 - CLAVATA1 homolog (AAN05447) from B. rapa, CsCLV1 - CLAVATA1 homolog (AAL77006) from C. sativa.

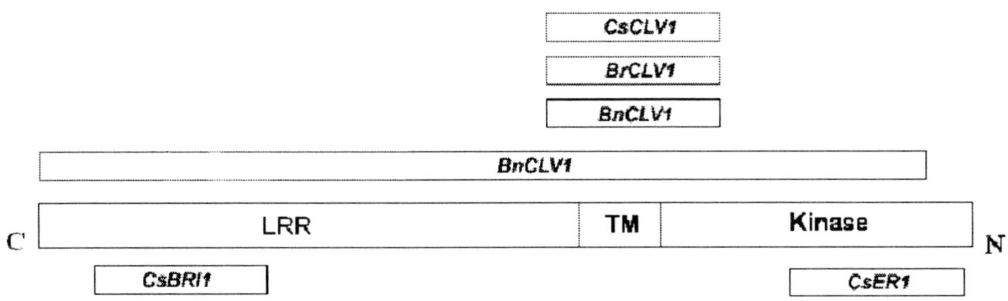

Fig. 2. Cloning the fragments of the receptor-like kinases of the Xa21 subfamily. The schematic representation of the cloned gene fragments regarding the LRR, transmembrane (TM), and kinase domains of the protein. CsBRI1 – BRASSINOSTEROID INSENSITIVE1 fragment from Camelina sativa; BnCLV1, BrCLV1 and CsCLV1 – CLAVATA1 fragments from Brassica napus, B. rapa, and; CsER1 – ERECTA1 fragment from C. sativa.

In arabidopsis, the *CLV1* mutations cause the accumulation of excess stem cells in the center of the SAM; as a result, internodes are shortened in characteristically fasciated shoots and inflorescences, with a concomitant increase in the number of leaves and bracts; other characteristic traits include extra flower organs, enlarged siliques, and higher seed numbers [6]. Such structural changes in shoots, inflorescences, flowers, and siliques of the Brassicaceae crop plants may represent the economically important traits and as such are under heavy artificial selection.

The characterized fragments of *Brassica* and *Camelina* genomes ([7], see Fig. 1) seem to represent the structural orthologs of the arabidopsis *CLV1*. The *CLV1* homologs within the Brassicaceae and outside this family primarily differ in the structure of the region corresponding to the transmembrane domain of CLAVATA1.

The radioactive probes based on the fragments of *BRI1, CLV1*, and *ER1* hybridized to DNA of both *Brassica* and *Solanum* plants (Fig. 3a). The *ER1* fragment corresponding to the most conserved kinase domain was the most efficient. The hybridization efficiency was lower with the short *CLV1* fragment corresponding to transmembrane plus kinase domains and especially with *BRI1* fragment corresponding to the LRR domain. The short *CLV1* probe was also effective in the RFLP analysis (Fig. 3b); so it will probably help establish the *CLV1* variation within the Brassicaceae and reveal the phenotypic manifestations of the *CLV1* control.

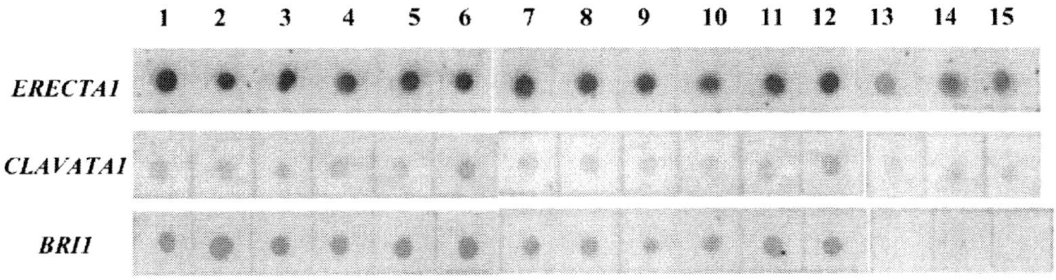

(a)

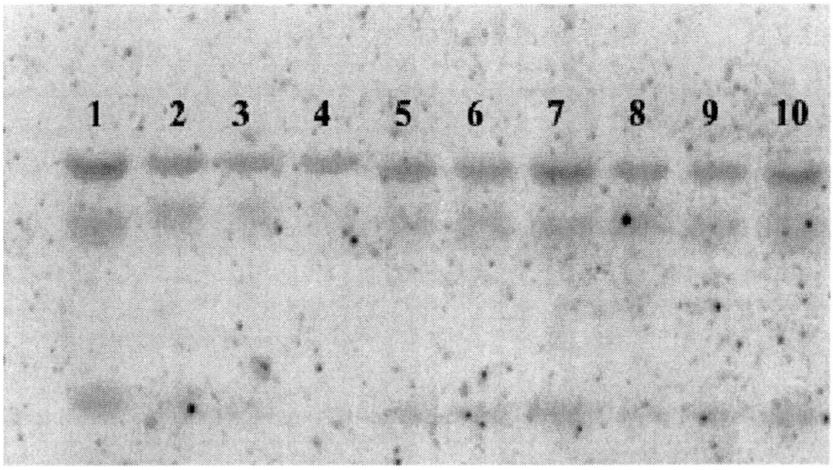

(b)

Fig. 3. Hybridization of genomic DNA samples from various Brassica species with the probes based on the homologs of arabidopsis ERECTA1 (ER1), CLAVATA1 (CLV1), and BRASSINOSTEROID INSENSITIVE1 (BRI1) genes of the Xa21 subfamily. Dot-hybridization. 1-6 – Brassica napus, 7-10 – B. rapa, 11, 12 – B. juncea, 13-15 –Solanum tuberosum; (b) RFLP analysis of a series of B. carinata genotypes.

In the search for the genes affecting the duration of the vegetative phase of plant development (earliness, days to flower), we cloned the fragments of the structural homologs of the genes *CONSTANS* (*CO*), *FLOWERING LOCUS C* (*FLC*), and *SHORT VEGETATIVE PHASE* (*SVP*), which control floral induction by photoperiod and vernalization. The properties of the cloned *CO*, *FLC*, and *SVP* homologs are listed in the table.

FLC was related to the span of life forms in the Brassicaceae, from spring annual to biennial plants [8]. However, not every *FLC* structural ortholog co-segregates with the days-to-flower trait [9]. To elucidate the physiological function of the *FLC* homologs isolated from *B. juncea* [10], we aligned the deduced amino acid sequences of the gene *FLC* and its structural homologs in arabidopsis and other Brassicaceae plants (Fig. 4). The function, rather than the origin from a particular plant species, was crucial for gene clustering. Of two *FLC* homologs isolated from *B. juncea*, one (AY266265) was orthologous to *FLC3* from *B. rapa*, whereas another homolog (AY268931), is probably derived from the genome of *B. nigrum*.

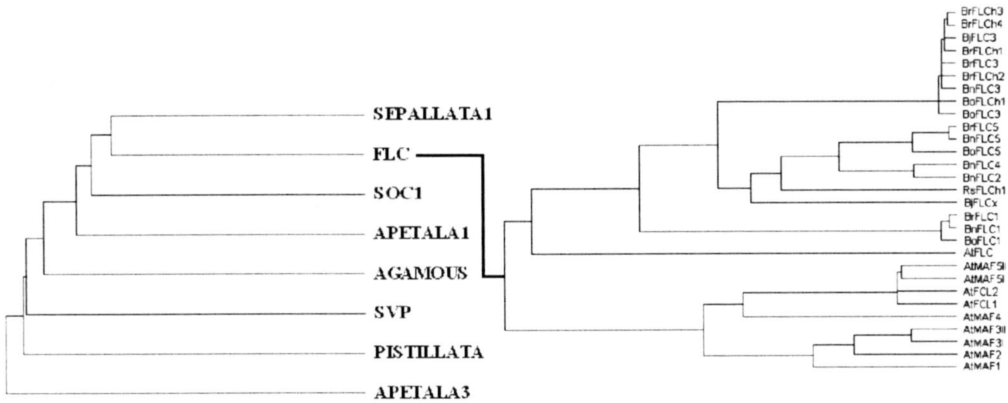

Fig. 4. The simplified phylogenetic tree of the arabidopsis MADS-box proteins. On the right, the comparison of the arabidopsis FLC prototype protein and its homologs corresponsing to exons 2--7 within the Brassicaceae family. AtFLC – Arabidopsis thaliana FLC (GenBank accession no. AAD21249); AtMAF1, AtMAF2, AtMAF3I, AtMAF3II, AtMAF4, AtMAF5I, and AtMAF5II – A. thaliana MAF (AAK37527, AAO65307, AAO65310, AAO65311, AAO65315, AAO65320, and AAO65321, respectively); AtFCL1 and AtFCL2 - FLC homologs from A. thaliana (AAG43854 and AAG43855); BnFLC - BnFLC5 - FLC homologs from Brassica napus (AAK70215--AAK70219, respectively); BrFLCh1 - BrFLCh4 and BrFLC1, BrFLC3, and BrFLC5 - FLC homologs from B. rapa (AAP31678--AAP31681 and AAO13159-AAO13157); BoFLCh1, BoFLC1, BoFLC3, and BoFLC5 - FLC homologs from B. oleracea (AAP31677, AAN87902, AAN87901, and AAN87900); RsFLCh1 - FLC homolog from Raphanus sativus (AAP31676); BjFLC3 and BjFLCx - FLC homologs from B. juncea (AAP42143 and AAP31243).

Dot-hybridization experiments demonstrated high activity of the probes derived from the *CO*, *FLC*, and *SVP*. The probe based on the potato homolog of *SUPPRESSOR OF OVEREXPRESSION OF CONSTANS1* (*SOC1*) also hybridized to *Brassica* DNA samples (Fig. 5). In addition, all four probes were shown to hybridize to *Solanum* DNA samples (data not shown). The *CO* probe was considerably less efficient than the probes derived from the MADS-box genes *FLC* and *SVP*.

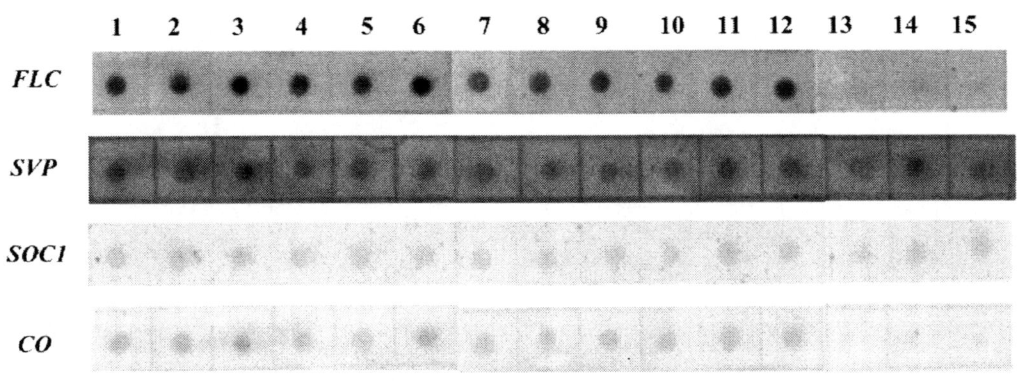

Fig. 5. Dot-hybridization of genomic DNA samples from Brassica and Solanum species with the probes based on the homologs of the days-to-flower genes FLOWERING LOCUS C (FLC), SHORT VEGETATIVE PHASE (SVP), SUPPRESSOR OF OVEREXPRESSION OF CONSTANS1 (SOC1) и CONSTANS (CO) from Brassica and Solanum. 1-6 – Brassica napus, 7-10 – B. rapa, 11, 12 – B. juncea, 13-15 – Solanum tuberosum.

A noticeable cross-hybridization activity with DNA and probes representing plant species from two dicot subclasses (Figs. 3 and 5) shows that the candidate-gene approach is instrumental in designing universal tools for the identification of the genes that control the key economically important traits of crop plants.

ACKNOWLEDGEMENTS

This work was supported by the Russian Foundation for Basic Research, project no. 01-04-48202. The authors thank P.E. Drobyazina for her help in cloning the *SOC1* homolog.

REFERENCES

[1] S. Pflieger, V. Lefebvre, M. Causse. The candidate gene approach in plant genetics: A review. *Mol. Breed.*, 2001, 7 (4): 275-291.

[2] A.E. Hall, A. Fiebig, D. Preuss. Beyond the arabidopsis genome: Opportunities for comparative genomics. *Plant Physiol.* 2002, 129 (4): 1439-1447.

[3] E.R. Morris, J.C. Walker. Receptor-like protein kinases: The keys to response. *Curr. Opin. Plant Biol.* 2003, 6 (4): 339–342.

[4] H. Schoof, M. Lenhard, A. Haecker et al. The stem cell population of *Arabidopsis* shoot meristems is maintained by a regulatory loop between the *CLAVATA* and *WUSCHEL* genes. *Cell*, 2000, 100 (6): 635-644.

[5] V.K. Sharma, J.C. Fletcher. Maintenance of shoot and floral meristem cell proliferation and fate. *Plant Physiol.*, 2002, 129 (1): 31-39.

[6] A. Dievart, M. Dalal, F.E. Tax et al. *CLAVATA1* dominant-negative alleles reveal functional overlap between multiple receptor kinases that regulate meristem and organ development. *Plant Cell*, 2003, 15 (5): 1198-1211.

[7] V.V. Martynov, I.L.Tsvetkov, E.E. Khavkin. Orthologs of the arabidopsis *CLAVATA1* gene in the cultivated Brassicaceae plants. *Russ. J. Devel. Biol.*, 2004, 35 (1): 00-00.

[8] S.D.Michaels, R.M. Amasino. Memories of winter: Vernalization and the competence to flower. *Plant Cell Envir.*, 2000, 23 (11): 1145-1154.

[9] M.E. Schranz, A.P. Quijad, Si.-B. Sung et al. Characterization and effects of the replicated flowering time gene *FLC* in *Brassica rapa*. *Genetics*, 2002, 162 (3): 1457-1468.

[10] V.V. Martynov, E.E. Khavkin E.E. Two Homologs of the *FLOWERING LOCUS C* Gene from Leaf Mustard (*Brassica juncea* Thell.), *Russ. J. Plant Physiol*, 2004, 51 (2): 00-00.

In: Biotechnology in Agriculture and the Food Industry ISBN 1-59454-119-1
Editor: G. E. Zaikov, pp. 77-79

Chapter 11

NEW BIOLOGICAL PREPARATIONS DESIGNED TO INCREASE CROP YIELDS AND REDUCE DOSES OF MINERAL FERTILIZERS

L. A. Sukhovitskaya
Institute of Microbiology,
National Academy of Sciences Belarus"
Kuprevich street 2, 220141, Minsk, Belarus
E-mail: soillab@mbio.bas-net.by Fax: (+375 172) 64-47-66

ABSTRACT

New biological preparations based on diazotrophic and phosphate – mobilizing microorganisms were elaborated to increase crop harvests and reduce doses of mineral fertilizers in Belarus.

Key words: biofertilizers, Phytostimophos, Rhizobacterin-C, diazotrophic and phosphate – mobilizing bacteria, plant growth stimulation.

Selection, characterization and application of microorganisms actively affecting crop yields are major theoretical and practical problems in soil and agricultural microbiology. Introduction of microbial agents into plant rhizosphere is widely recognized as a significant element of adaptive plant culture, especially in the areas critically depend on it.

Conclusive evidence obtained worldwide proves that rational use of nitrogen fixing bacteria is essential to intensify agropractice and introduce advanced technologies of supplying crops with atmospheric nitrogen taken up biologically.

The role of different diazotrophic microbial groups in soil nitrogen balance is not equivalent. In this respect bacteria of Rhizobium genus became the object of thorough investigation. Symbiosis of leguminous plants with nodulating bacteria provides for 10–90% of N requirements via nitrogen biofixation. Grain legumes are able to bind 10 to 150–200 kilos of nitrogen per hectare in a vegetative season, while perennial leguminous grasses fix up to 200–300 kg/ha. In comparison to grain cereals the root residues of grain legumes and perennial leguminous grasses contain elevated amounts of nitrogen– by 30–40 kilos/hectare and 120–200 kg/ha, respectively. The positive effect of legumes on subsequent crop yield is observed within next 3–5 years. Preliminary grain legume culture is equivalent to mineral fertilizer dose 50–70 kg N/ha, whereas perennial legumes are comparable to feeding 80–160 kilos N per hectare.

Adequate legume symbiosis with nodulating bacteria rules out administration of nitrogen fertilizers strategically, and in cases where biological nitrogen fixation appears insufficient for required harvests, the methods to intensify it should be developed.

The role of non – symbiotic diazotrophs was underestimated since their contribution into soil nitrogen stock was considered as minor. So far massive experimental findings have been piled on associative nitrogen fixation, methods of stimulation and application in perennial non – legume crops: wheat, barley, maize, rice, potato, fodder grasses and vegetables.

Generally, the scope of nitrogen fixation in non – legume associations is inferior to legume symbioses, ensuring 20 to 40% of non – legume N requirement. Yet, the potential of associative nitrogen fixation is rather high whereas microbial species responsible for the process are widely distributed and taxonomically affiliated to genera: Azospirillum, Azotobacter, Bejerinkia, Arthrobacter, Bacillus, Corynebacterium, Derxia, Enterobacter, Erwinia, Klebsiella, Vibrio and etc. Microbial agents would increase non – legume harvests and raise the content of biological nitrogen.

Apart from nitrogen, crop yields are limited by second significant nutrient–phosphorus. Overall P reserves in soil are relatively large reaching 10–20 ton per hectare. Regretfully, it is deposited in low – mobile water – insoluble state making it hardly available to plants. The bulk of administered P fertilizers interacts with ferrous and aluminium oxides (substances typical for sod – podzol Belarus soils) transforming into insoluble compounds. Phosphorus fertilizer efficiency does not exceed 25%. It is only natural that in Belarus where over 70% tillage is represented by sod podzol soils, and the manufacturer of P fertilizers – Gomel Chemical Plant cut the import of Russian feedstock and curtailed produce output, the problem of phosphorus deficiency has gained top priority. One of real keys to meet the challenge is microbial phosphate mobilization. It could result in release of 10 to 40% mobile and highly accessible to plants P_2O_5 from hardly soluble phosphates. Assessing the significance of microbial phosphate mobilization for crop production, it should be emphasized that real values of the process could be manifested only within rhizosphere zone where nutrient mobility gradient from soil to plant roots exists while practical efficiency of phosphate – mobilizing microorganisms is largely dependent both on phosphate – transforming capacity and the ability to colonize plant rhizosphere under competitive conditions with native and introduced microbial species.

Added to nitrogen – fixing and phosphate – mobilizing potential, almost all diazotrophs and some phosphate – mobilizing bacteria exert benevolent action by stimulating growth and photosynthesis in plants. It occurs due to several reasons:

(1) secretion of vitamins and phytohormones;
(2) production of pathogen inhibiting antibiotics;
(3) conversion of mineral elements into form readily available to plants.

Agricultural applications of microbial diazotrophic and phosphate – mobilizing preparations possessing a whole spectrum of valuable properties tend to broaden prospects as a consequence of aggravated environmental situation in the republic and disillusionment in technogenic yield – promoting means.

It should be admitted that contemporary agropractice in Belarus only to a limited extent engages plant – microbial interaction potential and the derived biopreparations which is mainly caused by lack of mass – scale production.

Rhizobacterin-C as an alternative to mineral nitrogen fertilizers is intended to increase grain crop harvests. Active principle is live culture and metabolites of associative diazotroph Klebsiella planticola. The strain displays such characteristics as fixation of atmospheric nitrogen, synthesis of β-IAA, suppression of root phytopathogens. It raises productivity at average rate 15% and saves nitrogen introduced under grain crops by 30-40kg/ha.

Phytostimophos alternative to mineral phosphorus fertilizers is designed for microbial phosphate mobilization, growth stimulation and rise of cultivar yields. Active principle – live culture and growth-promoting metabolites of Agrobacterium radiobacter. The strain carries out microbial transformation of hardly soluble phosphates in soil and fertilizers, stimulates seed germination, development and harvests of various agrocrops. It is equivalent in activity to 30-40 kg of P_2O_5 ammonized superphosphate.

Technologies of manufacturing biopreparations are launched at Bobruisk hydrolysis plant allowing to start in Belarus fabrication of bioproducts alternative to mineral fertilizers. The afore – mentioned biofertilizers have good application prospects in agronomy due to practical significance of biological N and P for national agriculture suffering from shortage of material resources and due to development of diazotrophic and phosphate – mobilizing microbial populations in plant rhizosphere contributing into maintenance of soil fertility in areas of intensive crop cultivation.

In: Biotechnology in Agriculture and the Food Industry
Editor: G. E. Zaikov, pp. 81-87
ISBN 1-59454-119-1

Chapter 12

DEVELOPMENT OF ISOLATED OVULES OF LENTIL *LENS ESCULENTA* MOENCH. *IN VITRO*

Galina N. Suvorova and Evgeniya A. Skotnikova
The All-Russia Research Institute of Legumes and
Groat crops, p/b Streletskoye, Orel, 302502, Russia
e-mail: office@vniizbk.orel.ru
fax: (0862)403130

ABSTRACT

There was developed a rescue technique of 5-19 days old lentil ovules, allowing the obtaining morphogenic tissue and rooted regenerated plants. The optimal content of growth regulators for induction of ovule development was at level BA (1…2,25) mg/l + IAA (0,2…1 mg/l) or BA (1…2,25 mg/l) + NAA (0,2…1 mg/l). A possibility of regenerated lentil shoots rooting after one-year subculturing was shown.

Key words: lentil, ovule, embryo, rooted shoots

INTRODUCTION

Culture *in vitro* of isolated fertilized lentil ovules *Lens esculenta* Moench. (*Lens culinaris* Medik.) first of all is important as ovule rescue technique, necessary for postgamic incompatibility overcoming by interspecific crosses. On the other hand the ovule culture can serve as a source of isolated lentil tissues obtaining.

A majority of researchers studying problems of lentil regeneration use either seedling fragments, or mature or nearly mature seeds as initial explants. A stimulating effect of cytokinin group – kinetin /1/, benzylaminopurine /2/, thidiazuron /3/ in shoot induction of primary lentil explants and inhibitory effect of benzylaminopurine on root development has been shown /4/. Plant regeneration from callus culture of lentil via somatic embryogenesis was observed /5/. A universal protocol based on seed culture was developed for efficient *in vitro* propagation of lentil /6/.

There are few reports available /7, 8/ on culture of immature ovules or embryos of lentil. The authors developed rescue technique for 14 days old embryos, having successfully applied the technique for interspecific crosses of *L. culinaris* with *L. ervoides*.

Available rich species genus *Lens* composition and the unused reserves of interspecific hybridization set a task of lentil embryo rescue technique improvement. The objective of this work was an elaboration of rescue technique for immature fertilized lentil ovules of various age.

Materials and Methods

The materials used were 5 - 19 days old fertilized lentil ovules of cv. Obraztsev Chiflik 7. The green pods were isolated from plants grown in the greenhouse. The immature ovules were excised from the pods and surface sterilized in 0,25% solution of chlorhexidini digluconate for 7 min. The ovules were then washed twice with sterile distilled water and placed on the medium. The induction media contained mineral salts of MS /9/ or B5 /10/ media, vitamins of B5 medium, 30 g/l of sucrose, 0,1 g/l of inositol, 1 g/l of hydrolysate casein, 5,5 g/l of agar supplemented with different concentration of growth regulators (Table 1). As growth regulators were used BA (0,2; 1; 2,25 mg/l), IAA (0,2; 1; 2 mg/l), NAA (0,2; 1; 2 mg/l). The rooting media contained mineral salts of MS medium, vitamins of B5 medium, 30 g/l of sucrose, 0,1 g/l of inositol, 2 g/l of glycine, 5 g/l of agar supplemented with growth regulators: IBA (1 mg/l), IAA (1mg/l) or NAA (1mg/l). For the first two days the explants were kept in darkness, then all cultures were maintained under 16-h photoperiod.

Table 1. Growth regulator composition and mineral base of culture media for lentil ovules

Medium	Growth regulators			Mineral salts
	BA	IAA	NAA	
B1	2,25	0,2	-	B5
B2	2,25	2	-	B5
B3	2,25	-	0,2	B5
B4	2,25	-	2	B5
B5	1	0,2	-	B5
B6	1	-	0,2	B5
B7	0,2	2	-	B5
B8	0,2	-	2	B5
M1	2,25	0,2	-	MS
M3	2,25	-	0,2	MS
M5	1	0,2	-	MS
M6	1	-	0,2	MS
M7	0,2	2	-	MS
M8	0,2	-	0,2	MS
M9	1	1	-	MS
M10	1	-	1	MS

RESULTS

Inoculated to the nutrient media isolated lentil ovules either necrotized, or continued their development *in vitro*. Ovular integuments dieing and embryo uncovering accompanied the development. As a rule in the presence of growth regulators embryos developed abnormally with induction of multiple buds and shoots and lack or rudimentary development of a rootlet (Fig. 1). In single cases, the callus was formed out of embryo tissues.

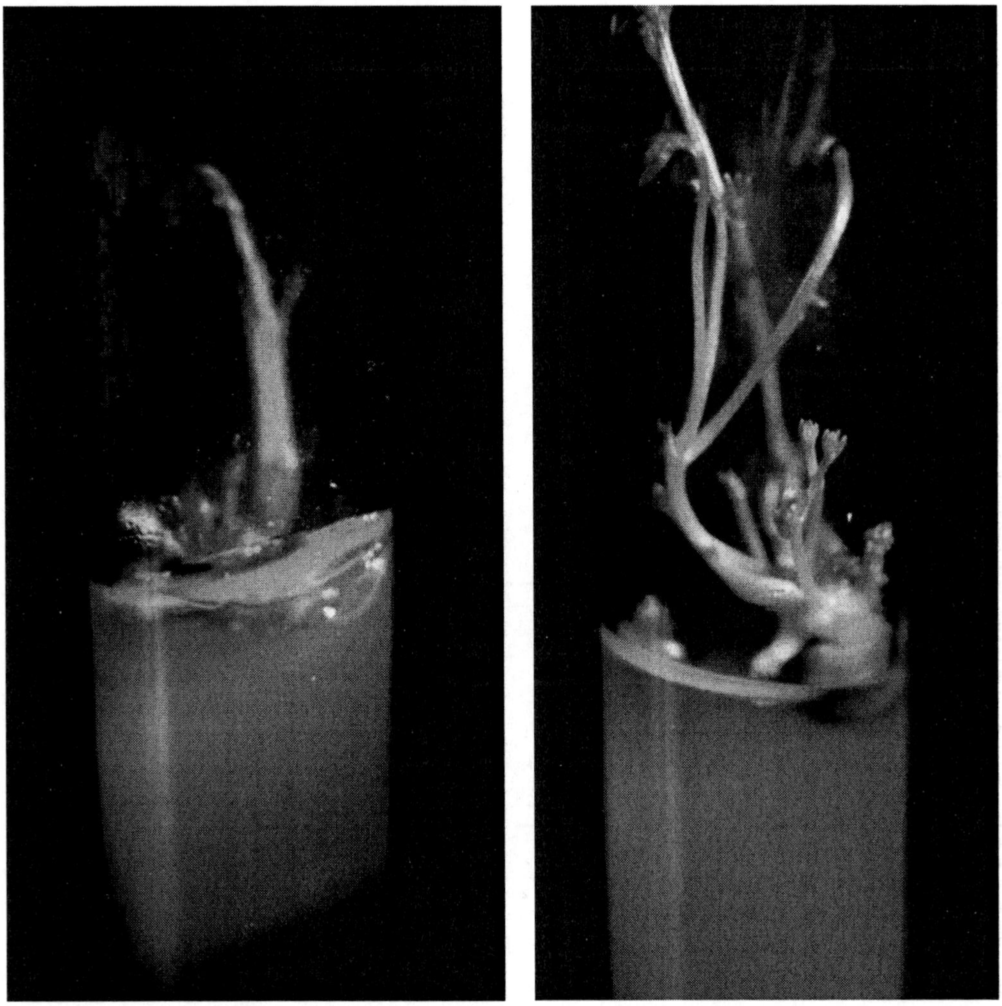

A B

Fig.1. Development of immature ovules of lentil on induction media: A - 8 day old ovule; B – 18 day old

The survival rate of embryos went up with the increase of their maturity degree (Table 2). In group of 5-7 diurnal embryos 16,7 % of explants had ability for self-contained development *in vitro*. At the age of 18 - 19 days 81,8 % of embryos survived on nutrient media.

Table 2. Development of lentil ovules *in vitro* depending on their age

Ovule age, days	No. ovules inoculated	Way of development			Survival,%
		abnormal growth	callus	necrosis	
5-7	48	7	1	40	16,7
8-9	116	48	2	66	43,1
10-11	111	47	4	60	45,9
12	81	56	3	22	72,8
14-15	88	52	2	34	60,2
18-19	33	27	0	6	81,8

Embryo survival rate at various media varied from 6,7 up to 71,1% (Table 3). Media used in experiment differed by composition of mineral salts and quantitative ratio of auxins and cytokinins. The most efficient for embryo development were media with mineral base MS in comparison to media keeping mineral salts B5. On mediums with MS salts a survival rate of ovules varied from 38,5 up to 71,2 %. Proceeding from parameters of the survival rate, the optimal content of growth regulators for an induction of immature lentil ovules development balanced at a level BA (1…2,25) mg/l + IAA (0,2…1 mg/l) or BA (1…2,25 mg/l) + NAA (0,2…1 mg/l). In single cases on media with NAA formation of callus tissue was observed.

Table 3. Development of immature lentil ovules *in vitro* depending on medium composition

Medium	No. ovules inoculated	Way of development			Survival,%
		abnormal growth	callus	necrosis	
B1	15	1	0	14	6,7
B2	14	2	0	12	14,3
B3	12	2	0	10	16,7
B4	14	0	3	11	21,4
B5	10	4	0	6	40,0
B6	9	2	0	7	22,2
B7	8	4	0	4	50,0
B8	7	2	2	4	57,1
M1	49	25	0	24	51,0
M3	57	33	1	23	59,6
M5	78	41	0	37	52,6
M6	74	45	3	26	64,9
M7	14	8	0	6	57,1
M8	13	4	0	9	38,5
M9	51	30	0	21	58,8
M10	52	34	3	15	71,2

Developing out of the ovule abnormal seedling, as a rule already with cluster of buds or shoots, was passed to the medium MS, containing 0,6 mg/l BA. On the given medium there was an increase of mass of morphogenic tissue and elongation of shoots (Fig. 2). Morphogenic tissues were subcultured on the same medium, alternating with some variants of inductive media. For the best developing of shoots, the tissues were passed to the hormone free MS medium.

Fig.2. Morphogenic tissue of lentil derived from embryo

Developed on MS medium or on subculture media regenerated shoots were cut off from their forming tissue and transferred to rooting media. For rooting of regenerated shoots were used 3 variants of rooting media, containing as auxins IBA, IAA or NAA. Root shaping of regenerated shoots took place on all variants of media, with rooting efficiency varied from 9,1 up to 42,8 % (Table 4). On media keeping IBA or IAA well developed roots were formed, on medium with NAA there was a growth of dwarfed roots, frequently with callus formation; callus was formed also without root development (Fig. 3). With increase of stay duration in culture *in vitro* rooting efficiency decreased. After year of subculturing it was much more difficult to root regenerated lentil shoots than in the first months of their growth.

Shoots which have generated little rootlets, were extracted from medium, the roots were washed under water and kept in glass tubes with water for two-days. Then the plantlets were transferred to pots with soil into a greenhouse. From 24 planted in the greenhouse plantlets, 10 were successfully acclimatized *in vivo*. The percentage survival of the plantlets after transfer to soil was 41,7 %. What is more the survival rate after rooting on media with IBA or IAA was higher then that on medium with NAA. Regenerated plants normally developed in the greenhouse, blossomed and formed pods and mature seeds. Distinct morphological differences at regenerated plants in comparison to initial cultivar were not revealed.

Table 4. Rooting response of regenerating lentil shoots on different media after 2, 6 and 12 months of subculture

Subculture period	Growth regulator composition, mg/l	No. shoots planted	No. rooted shoots	Rooting efficiency, %
2 months	IBA-1	14	5	35,7
	IAA-1	11	1	9,1
	NAA-1	9	2	22,0
6 months	IBA-1	19	3	15,7
	IAA-1	14	6	42,8
	NAA-1	22	4	18,2
12 months	IBA-1	32	5	15,6
	IAA-1	30	3	10,0
	NAA-1	27	1	3,7

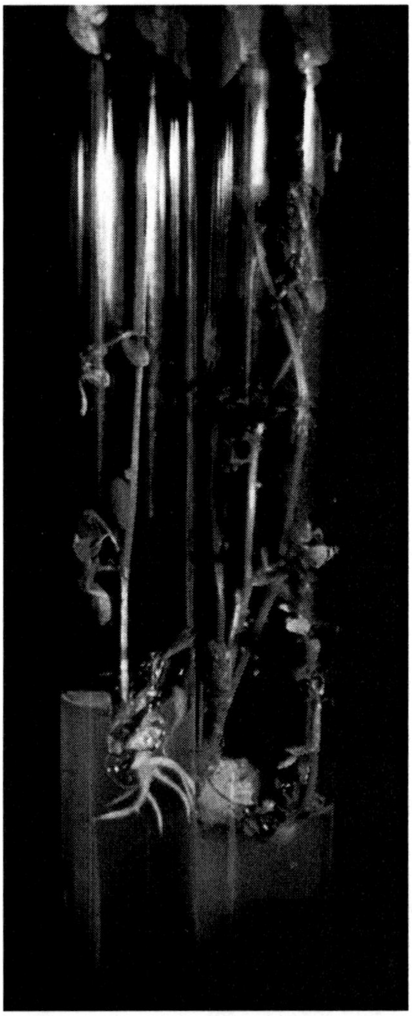

Fig.3. Root growth (left) and callus formation (right) in regenarated lentil shoots on different rooting media

DISCUSSION

Having cultivated isolated immature lentil ovules on media in the presence of cytokinins and auxins, we obtained morphogenic tissue, suitable for long-term culture *in vitro* and formation of multiple regenerated shoots. We showed an ability of independent isolated lentil ovules development starting from 5 days after fertilization. Previously known lentil embryo rescue technique /7/ is developed only for 14-day age and assumes germination of an embryo on the medium with formation of a single plant. Our developed method can simultaneously be used as embryo rescue technique and as a way of their micropropagation.

The problem of regenerated lentil shoots rooting *in vitro* has not been solved for a long period /1, 2, 11/. Successful results were obtained in rooting of regenerant shoots (4.6 – 39.9%) on a medium containing 2 mg/l of IAA after 8 weeks of culture /4/. High rooting ability (more than 50%) of shoots produced from seeds on a medium supplemented with 1,5 mg/l of NAA was shown /6/. We have demonstrated a possibility of rooted regenerated lentil plants obtaining after one year of subculturing. Efficiency of rooting was relatively low on all media. Probably root induction of regenerated lentil shoots to a greater extent is influenced by other unexplored factors than by medium composition.

Thus as a result of the investigations a rescue technique for immature lentil ovules of 5-19 days age, permitting to obtain morphogenic tissues and rooted regenerated plants is developed. The possibility of self-contained *in vitro* development of isolated lentil ovules beginning with 5-day age is shown.

REFERENCES

[1] D.J. Williams, A.Mchugen: Plant Cell, Tissue and Organ Culture, 7, 149 (1986)

[2] M.C. Polanco, M.I. Pelaez, M.L. Ruiz: Plant Cell, Tissue and Organ Culture, 15, 175 (1988)

[3] K.A. Malik, P.K. Saxena: Aust. J. Plant Physiol, 19, 731 (1992)

[4] M.C. Polanco, M.L. Ruiz: *Plant Cell Reports*, 17, 22 (1997)

[5] P.K. Saxena, J. King: *Plant science*, 52, 223 (1987)

[6] G. Ye, D.L. McNeil, A.J. Conner, G.D. Hill: *New Zealand Journal of Crop and Horticultural Science*, 30, 1 (2002)

[7] D. Cohen, G. Ladizinsky, F.J.Muehlbauer: *Plant Cell, Tissue and Organ Culture,* 3, 343 (1984)

[8] G. Ladizinsky, D. Cohen, F.J.Muehlbauer: *Theor. Appl. Genet*, 70, 97 (1985)

[9] T. Murashige, F. Skoog: *Physiol. plant*, 15(13), 473 (1968)

[10] O.L. Gamborg, R.A. Miller, K. Ojima: *Exp. Cell Res*, 50(1), 151 (1968)

[11] V. Kumar, B.M. Singh, S. Singh, P.K. Rathore: *4th European Conference on Grain Legumes*, Cracow, 148 (2001)

In: Biotechnology in Agriculture and the Food Industry ISBN 1-59454-119-1
Editor: G. E. Zaikov, pp. 89-94

Chapter 13

OBTAINING OF OSMORESISTANT CALLUS LINES OF SOWING PEA (*PISUM SATIVUM* L.)

Galina Soboleva
All-Russia Research Institute of Legumes and Groat Crops,
p/b Streletskoye, Orel, 302502, Russia
e-mail: office@vniizbk.orel.ru
fax*: (0862)403130*

ABSTRACT

The analysis of callus pea lines growth on selective media with various content of polyethylene glycol (PEG 6000) for selection of osmoresistant genotypes in culture *in vitro* is carried out. Low PEG concentrations (5-10 %) rendered stimulating influence on growth and development of callus tissues. At strengthening PEG concentration up to 30 % and duration of subculturing the number of growing calli sharply dropped. The lethal concentration of the selective factor amounting to 35 % PEG is specified. Callus pea lines, resistant to PEG in concentration of 15-20 % during six passages are obtained.

Key words: *Pisum sativum* L., callus, polyethylene glycol.

INTRODUCTION

An actual problem of sowing pea breeding is raising drought-resistant cultivars. Cell selection inclusion as one of raising stages of drought-resistant cultivars will allow speeding up this process. It is possible to model water deficiency *in vitro*, using selective systems with osmotically active substances. It is shown, that polyethylene glycol (PEG) in composition of nutrient medium can serve as the adequate selective agent reflecting reaction of the whole plants on a drought /1, 2/. Reports on cell selection for osmoresistance in culture *in vitro* of pea are not known. The objective of our work was in a sampling procedure development of callus pea lines in culture *in vitro*, resistant to osmotic stress.

MATERIALS AND METHODS

Long-term passed morphogenic callus tissues pea genotypes Orlus and Az 93-1964, introduced into culture in 1996 ... 1997 served as a material for selection of resistant to osmotic stress pea lines. Orlus is a leafless, middle-ripening type cultivar, with the leaf modified in strongly developed tendrils. Az 93-1964 is a new form of pea "chameleon", obtained in All-Russia Research Institute of Legumes and Groat Crops, and distinguished by canopy foliar heteromorphism /3/.

Medium MS_{12}, containing mineral salts of MS medium /4/, vitamins of B5 medium /5/, 30 g/l sucrose, 100 mg/l inositol, 5 mg/l 6-benzylaminopurine (BA) and 2 mg/l naphthylacetic acid (NAA) was used as a control one. Polyethylene glycol (PEG) with molecular weight 6000 in concentrations from 10 up to 40 % was used as a selective factor. The media containing polyethylene glycol with molecular weight 6000, remain liquid, therefore callus tissues were incubated on filter paper substrates. Scraps of callus tissues with weight 40 ... 50 mg served as initial explants. Calluses were cultured at light intensity of 2000 lk fluorescent lamps and 16-h photoperiod. The gain of callus tissues in selective conditions was estimated by augmentation of fresh weight. Magnitude of gain was expressed relatively to initial weight in percentage.

RESULTS

Media with different content of polyethylene glycol were used for deriving of resistant to osmotic stress callus pea lines (Table 1). 1760 initial callus explants of peas genotypes Orlus and Az93-1964 were planted on selective media. In controlled conditions (without PEG) callus tissues of the investigated genotypes grew practically equally. Growth of callus tissues was rather fast; callus had intensive green color with beginning of shoot organogenesis.

Table 1. Nutrient media composition during selection of callus pea lines for osmoresistance.

Media	Composition
MS_{12} (St)	Mineral salts MS, vitamins B5, sucrose 30 г/л, inositol 100 mg/l, BA5 mg/l, NAA 2 mg/l
1	St + 10% PEG
2	St + 15% PEG
3	St + 20% PEG
4	St + 25% PEG
5	St + 30% PEG
6	St + 35% PEG
7	St + 40% PEG

Growth activity of pea callus tissues on selective media with PEG depended on concentration of selective factor and on genotype of used material (Fig. 1). At 10 % of PEG concentration in nutrient medium there was an augmentation of gain of callus pea weight in comparison to standard medium. Reaction of genotypes Orlus and Az 93-1964 on presence of

10 % PEG in medium was the same. The gain of callus tissue at the given genotypes has exceeded parameters of callus weight gain on standard medium by 7...8 %. In this case polyethylene glycol in 10 % of concentration did not in the least suppress regeneration, but on the contrary, promoted morphogenesis processes. Reaction of callus tissues on increasing of PEG concentration in medium was expressed in retardation of growth rates and depressing of regeneration processes intensity. At strengthening PEG up to 35 and 40 %, callus growth was not observed; there was a gradual decolorization of tissues and necrosis. During the first passage, death of all inoculated explants of both genotypes was observed. Thus 35 % polyethylene glycol concentration in selective medium is lethal for pea callus tissues.

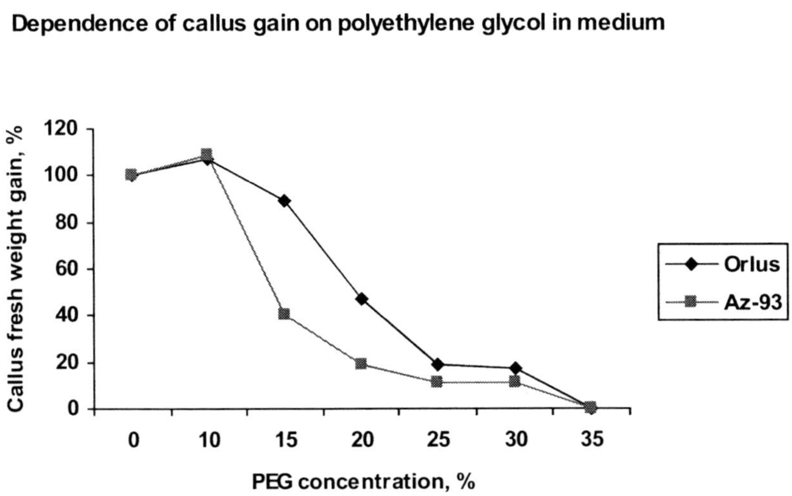

Fig.1

For selection of cells, truly tolerant to osmotic stress, principle of subculture was used at the presence of selective factor during six passages. Passage duration went on for 45 days. For comparative evaluation of various PEG concentrations influence on growth of pea callus, a survival rate parameter of callus explants number was estimated at the end of everyone passage (Tab. 2).

At long-term culture in the presence of 15 % PEG the number of growing callus genotypes Orlus and Az 93-1964 decreased insignificantly. Practically all the selected clones in the first passage were maintained to the seventh passage. At strengthening PEG concentration up to 20 % and prolonged subculturing the number of growing callus sharply dropped. By the end of the sixth passage the survival rate of callus cultivar Orlus had made 30 %, genotype Az 93-1964 - 14 %. Especially sharp elimination of callus took place on media with high content of selective factor. In the presence of 25 % PEG in nutrient medium the survival rate of callus cultivar Orlus in the fifth passage decreased up to 2,5 %. Similar reaction was also observed at 30 % PEG concentration. As the subculturing went on during the sixth passage there was necrosis and death of all callus clones. Genotype Az 93-1964 during subculturing on selective media was characterized by the greater sensitivity to PEG. On media with 25-30 % PEG content, callus viability was preserved only during three passages.

Table 2. Survival rate of callus pea explants during subculturing on selective media with PEG-6000, %

Medium	Inoculated explants	Orlus						Az 93-1964					
		passage						passage					
		1	2	3	4	5	6	1	2	3	4	5	6
St	130	100	100	100	100	100	100	100	100	100	100	100	100
St+10% PEG	120	100	-	-	-	-	-	100	-	-	-	-	-
St+15% PEG	120	100	100	99,2	99,2	99,2	95,0	100	100	90,0	90,0	90,0	84,2
St+20% PEG	120	98,3	99,2	97,5	65,0	41,2	30,0	87,5	70,5	27,5	20,0	18,3	14,2
St+25% PEG	120	94,2	90,3	63,3	15,8	2,5	0	79,2	54,7	12,5	5,0	0,8	0
St+30% PEG	120	93,3	54,5	10,0	5,8	2,2	0	69,2	15,7	9,2	0	0	0

After three passages on selective media callus clones divided into two parts. One was used for further culturing in selective media, another - for induction of regenerating processes and deriving of plants (Fig. 2).

Selection scheme of osmoresistant pea calli

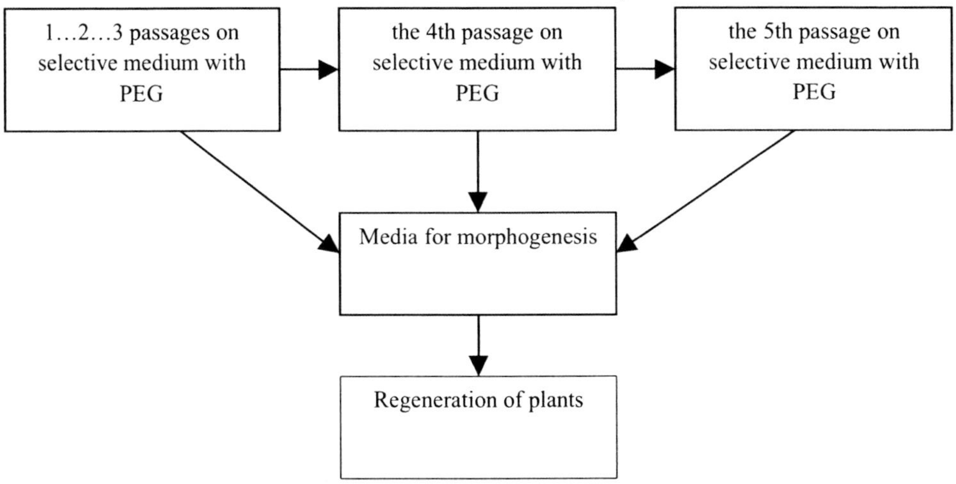

Fig. 2

An induction of regenerating processes in the survived callus clones was carried out in two steps. At the first stage there was an increase of callus masses and induction of stem morphogenesis in unorganized callus culture. For this purpose callus were transferred to agarized medium MS_{12} containing 5 mg/l BA and 2 mg/l NAA. Medium MS_{12} was used for callus tissues mass augmentation, as primarily small inoculums were cultivated on selective media. For obtaining of regenerated shoots the callus were transferred to medium MS containing 5 mg/l BA and 0,2 mg/l NAA.

At the second stage, the developed regenerated shoots were transferred on medium B5 for rooting with concentration of the basic components diminished twice as little and supplemented with 1,5 mg/l of NAA. Rooted regenerated plants R_0 genotypes Orlus and Az 93-1964 were obtained from callus clones, selected during three passages on nutrient medium, containing 15 % PEG.

DISCUSSION

Experience in use of cell selection methods with application of osmoticum during breeding of drought-resistant plants is still very little. Plants, tolerant to polyethylene glycol - induced water stress, were obtained from lucerne cell line /6 /; resistant lines of tomatoes /1/ were selected on media containing polyethylene glycol. Regenerated plants, which gave property of water stress resistance to seed generation, were obtained from PEG osmotic resistant cell lines of rice /7/, maize /8/. Techniques of somatic cells selection of sorghum, resistant to osmotic stress are being developed /2/.

The outcome of our experiments on pea shows, that a selective system with polyethylene glycol can be used for selection of callus lines for osmoresistance. The survival rate and growth callus rates depend on PEG concentration, duration of culturing on selective media and on genotype. The lethal concentration of the selective factor amounting to 35 % PEG is specified. Low PEG 10% concentration renders stimulating effect on growth and development of callus tissues. Ability of pea callus to induction of regenerated shoots after subculturing on selective media with low osmotic stress is shown.

REFERENCES

[1] R.A. Bressan, P.M. Hasegawa, A.K. Handa: *Plant Science Letters*, 21, 23 (1981).

[2] R.H. Smith, S. Bhaskaran, F.R. Miller: *In Vitro Cell. Develop Biology*, 21 (10), 541 (1985).

[3] A.N. Zelenov, A.V. Amelin, N.E. Novikova: *Rep. Russ. Acad. Agric. Sci.*, 4, 15 (2000). (in Russian).

[4] T. Murashige, F. Skoog: *Physiol. Plant.*, 15 (13), 473 (1962).

[5] O.L. Gamborg, D.E. Eveleigh: *Can. J. Biochem*, 46 (5), 417 (1968).

[6] A.I. Atanassov: *Biotechnology in Plant Growing*, Sofia, 1988, 245 p.

[7] S.L. Belyanskaya, Z.B. Shamina: *Physiol. Plant*, 40 (4), 681 (1993). (in Russian).

[8] Yu. I. Dolgykh, S.M. Larina, Z.B. Shamina: *Physiol. Plant*, 41 (1), 114 (1994). (in Russian).

In: Biotechnology in Agriculture and the Food Industry ISBN 1-59454-119-1
Editor: G. E. Zaikov, pp. 95-100

Chapter 14

EXPRESS-APPRAISAL OF PLANT PROTOPLAST NATIVITY

V. A. Kudryashova and A. P. Kudryashov
Belorussian State University, Biological faculty
Scoryna av., 4, Minsk, Belarus,
e-mail: verakudryashova@mail.ru, fax: (375-17) 277-55-35

ABSTRACT

Influence of treatment conditions to the isolation of native protoplasts from the leaves of 5 plant species was investigated. During the experiments optimal conditions for the isolation of intact plant protoplasts were elaborated. It was shown that the most acceptable method of express-appraisal of protoplast nativity is method of deplasmolysis. Methods of treatment by vital colorants in a number of cases can't be used. The evident correlation between the results of express-appraisal of protoplast nativity and ability of protoplasts obtained to proliferation was shown.

Key-words: protoplasts, plasma membranes, estimation of nativity, proliferation

Plant protoplasts are widely used in present-day biotechnology. Influence of auxesis, antibiotics and toxins to plant cells can be studied by using protoplasts. With the advent of methods which permit to isolate plant protoplasts, plasma membrane of plant cells became easily accessible for direct observations. Investigation of electrophysiological processes on the membrane surface, processes of transport, influence of antibiotics, cell toxins and other bioactive substances directly to plasma membrane can give useful information not only for plant physiology, but for human physiology, biochemistry, biophysics and other sciences.

At present time the most promising branch of biotechnological science with using plant protoplasts is genetic modification of plants. It means that we have to insert new genetic material into plant protoplasts and than regenerate whole plants from such modified protoplasts. Different plant genes, animal and bacterial genes are transferred to plant protoplasts [1]. Insertion of whole organelles (chloroplasts) into protoplasts allows to increase of crop productivity on a large scale [2]. There are some scientific works by insertion of cyanobacteria cells into plant protoplasts.

Protoplast fusion is widely used in plant selection. In such works scientists unite protoplasts of genetically distant plants, and sometimes they unite protoplasts with "cytoplasmic droplets", i.e. fragments of cytoplasm, which are surrounded by plasma membrane. Such "droplets" contain organelles, but devoid of nuclei.

For getting of reliable and reproductible results, experimenter must be sure in nativity of protoplasts obtained. There are set of express-methods of protoplast nativity appraisal. Most of them are based on the investigation of their plasma membrane state. These are method of deplasmolysis in hypotonic media, method of treatment of vital colorants, but the most reliable appraisal of plasma membrane state can give the investigation of it's electrophysiological characteristics [3, 4, 5].

But maintenance of plasma membrane parameters at the normal level doesn't signify that intracellular metabolic processes wasn't dislocated during the procedure of protoplast isolation, and such protoplasts are able to regenerate of cell wall and proliferate.

The aim of this work was the elaboration of optimal conditions of isolation of protoplasts of 5 plant species, as well as definition of application of express-appraisal methods on the basis of testing of protoplasts obtained to proliferation ability.

Materials and Methods

Protoplasts was obtained from the young leaves of 5 plant species (*Drimiopsis maculata, Begonia Feastii, Setcreasea purpurea, Aloe jucunda* and *Nicotiana tabacum)* by treatment of elaborated enzyme preparation, which contains pectinase (0,5 U/mg) and cellulase "Onozuka R-10" (1,8 U/mg) in concentrations 0,5 % and 2 % respectively. Composition of enzyme preparation was selected on the ground of the results of each enzyme influence to the rate of cyclosis in internodal cells of charophyte alga *Nitella flexilis* (cytokinetic testing).

Solution for protoplast isolation contained mannitol (400mM) as osmolytic agent. pH value of the solution was maintained up-to-date 5,6 by buffer mixture of MES-Tris.

Optimal conditions for protoplast isolation were selected empirically by variation of enzyme preparation composition, temperature and continuation of exposure of leaf fragments in enzyme solutions particularly for each plant species.

All stages of protoplast isolation was made in aseptic conditions. *Nicotiana* plants was initially aseptic, because they were grown in sterile nutrient medium. The leaves of other plants was sterilized by bleach, containing sodium superoxide with the concentration of reacting substance 2 %. Continuation of leaf fragment treatment by solution of sterilizing agent was 15–20 min for the leaves of *Begonia* and *Aloe*, and 10–15 min for the leaves of other plants. Then, sterilizing substance was washed away from the surface of leaf fragments by sterile water. Selection of the sterilizing agent was based on the results of cytokinetic testing of different sterilizing substances on *Nitella flexilis.*

Protoplasts obtained were washed out by medium, which contained some components, which can maintain the viability of protoplasts during the experiment. This medium had pH value on the level of about 7,5–8,0 to inactivate the rests of enzymes.

Protoplast nativity was estimated by treatment of neutral red and fluorescein diacetate, as well as by protoplast swelling ability in hypotonic media. In a number of cases protoplast nativity can be estimated by sight: by viscosity of solution for isolation, by protoplast shape and spacing of chloroplasts etc.

Proliferation ability was investigated only on the protoplasts obtained from the haploid plants of *Nicotiana tabacum*. Protoplasts were placed on the surface of dense nutrient medium MS with 2,4-D and kinetin in concentrations 0,25 and 1,25 mg/l respectively. With the object of arrangement of optimal osmotic conditions sucrose concentration in MS medium was enhanced from 30 to 100 g/l. Protoplasts were cultivated during 6 weeks. Cultivation temperature was 24,5 °C.

Microscope with the add-in digital camera was used for the fixation of the experiment results.

Results

During the experiments optimal conditions for protoplast isolation from the leaves of 5 plant species were selected. Incubation temperature was 6 °C or 24 °C and continuation of incubation varied from 4 to 48 hours. The results obtained are shown in the table 1.

Table 1. Characteristics of protoplasts, isolated from the leaves of plants under the different incubation conditions

Plant	Optimal conditions for protoplast isolation	Nativity of protoplasts obtained, %	Average protoplast yield, N/sm^2
Drimiopsis maculata	24 ч, 6°C 4 ч, 24°C	86,7±6,0 79,4±12,0	10263 28120
Begonia Feastii	48 ч, 6°C	70±6,5	70735
Setcreasea purpurea	4 ч, 24°C	47,2±10,1*	9814
Aloe jucunda	4 ч, 24°C	75,5±4,0	—**
Nicotiana tabacum	12 ч, 6°C	70±3,5	10560

*neutral red was sorbed by 75,5±30,4 % cells

**we didn't define protoplast yield for *Aloe*, because in this case protoplasts were isolated from the inner mesophyll tissues

This table shows optimal conditions for isolation of protoplasts in the sufficiently high yield of native (intact) protoplasts. Increasing of continuation of exposition in a number of cases led to increasing of protoplast yield, but in this cases strong decrease of degree of nativity was observed.

Comparative analysis of the methods of express-appraisal of protoplast nativity showed, that the most convenient method is method of deplasmolysis. The method of neutral red colouring is more visual, but sometimes it can't be applied, because some plant cells contain natural red pigments (anthocyanins). On the whole, the results, obtained by using of both methods are coincide, but in one case (*Setcreases purpurea*) the great differences between the results were noticed (see Table). Moreover, it was demonstrated, that some fragments of destroyed tissues sorbed the colorant more intensive, than intact protoplasts.

Method of fluorescein diacetate treatment demands of close washing the protoplasts from the enzyme rests as well as cell debris to leave out the action of etherases in extracellular solution.

In a number of cases (*Drimiopsis maculata)* extent of protoplast damage could be defined to a great approximation without any treatment. Intact protoplasts had round or oval shape, chloroplasts were distributed evenly in whole volume of cytoplasm. Protoplasts with any damages had irregular ameboid shape. Chloroplasts were aggregated and accumulated at one of the poles of such protoplast (Fig. 1). In this case enzymatic medium after plant material incubation became viscous and contained a great number of chloroplasts.

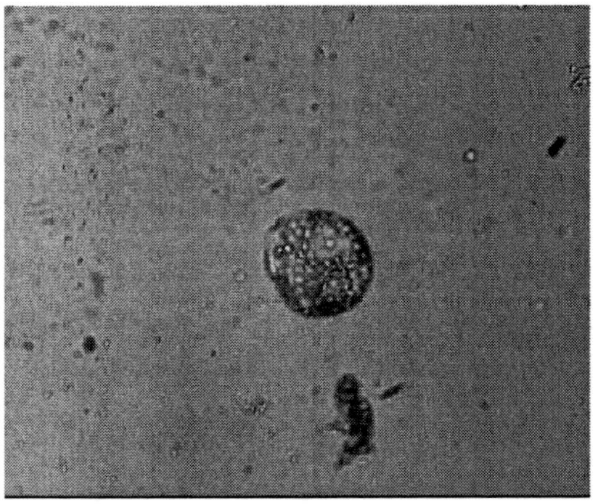

(a)

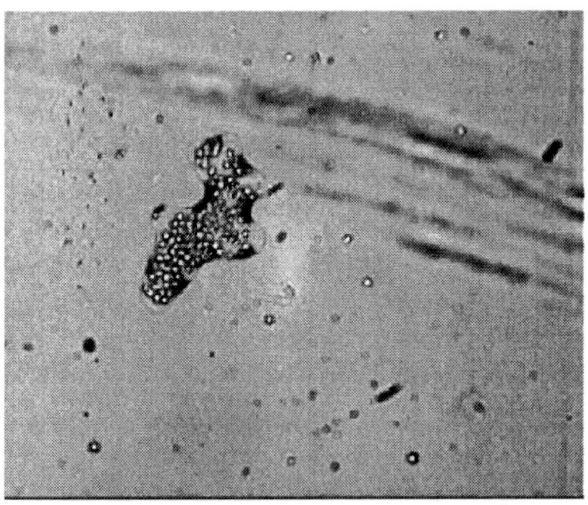

(b)

Fig. 1. Protoplasts, isolated from the leaves of *Drimiopsis maculata*. a – intact protoplast; b – damaged protoplast

Express-methods analysed can't give exact information about preservation of intracellular metabolic processes, because they evaluate only state of plasma membrane. With the object of the establishing the state of intracellular processes in protoplasts obtained, we have investigated their ability to cell wall regeneration and proliferation. These investigations were made by using of tabacco protoplast suspension diluted to 200 cells/ml and than was placed to the surface of dense nutrient medium. So, individual protoplasts were separated from one another.

In 6 weeks of cultivation, microcolonies of small shapeless cells appeared on the surface of the nutrient medium. This microcolonies could be examined only under the microscope (Fig. 2). Evidently, each colony was the result of proliferation of one protoplast. Practically 100 % of microcolony cells sorbed neutral red. It was the evidence of their viability.

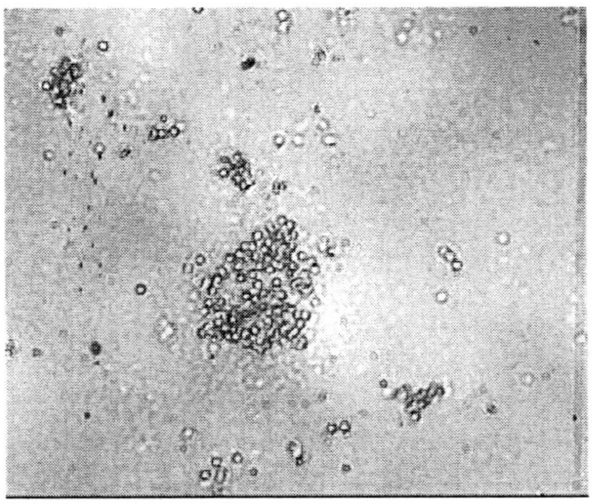

(a)

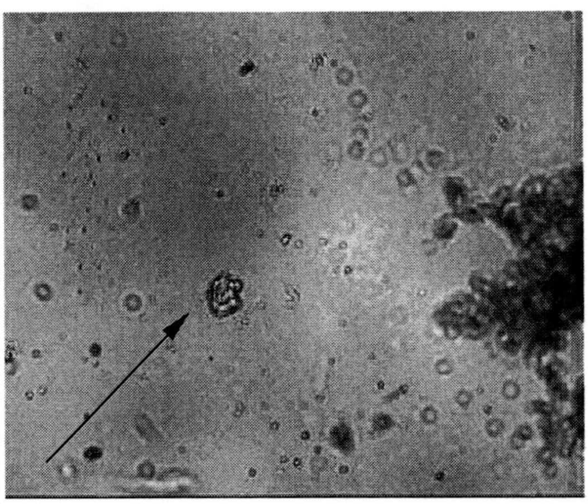

(b)

Fig. 2. Protoplast proliferation on dense nutrient medium after 6 weeks of cultivation a – the microcolony of *Nicotiana tabacum* cells (50x magnificatuon); b – idem microcolony after neutral red treatnent. Arrow shows the single cell (100x magnification)

So, there is an evident correlation between the results of express-appraisal of protoplast nativity and the ability of protoplasts to cell wall regeneration and proliferation.

REFERENCES

[1] Delbarre A., Muller P., Imhoff V., Barbier-Brygoo H., Maurel C., Leblanc N., Perrot-Rechenmann C., Guern J.: *Plant Physiology*, v. 105, N 2 (1994), pp. 563 – 569

[2] Beker M., Kliepinysh G., Rajpulis E.: *Biotechnology*, 1990, M

[3] Abe S., Takeda J.: *Journal of Experimental Botany*, v. 37, N 175 (1986), pp. 238 – 252

[4] Potossin I., Andjus P.: *Plant Physiology*, v. 106 (1994), pp. 313 – 319

[5] Abe S., Takeda J., Seneda M.: *Plant & Cell Physiology*, v. 24, N 4 (1980), pp. 537 – 546

In: Biotechnology in Agriculture and the Food Industry ISBN 1-59454-119-1
Editor: G. E. Zaikov, pp. 101-107

Chapter 15

DEVELOPMENT OF MICROBIAL PESTICIDES

E. I. Kolomiets and A. G. Lobanok
Institute of Microbiology, National Academy of Sciences, Belarus
Kuprevich str.,2, 220141 Minsk, Belarus
E-mail: kolomiets@mbio.bas-net.by, fax: 0037517 264 47 66

ABSTRACT

Results of basic and applied studies conducted at Institute of Microbiology, Belarus National Academy of Sciences to elaborate biological agents controlling pathologies and pests in plant cultivars have been presented. Development into pilot plant production of microbial preparations in Belarus was demonstrated.

Key words: antagonist, *Bacillus subtilis,* biocide, entomopathogen, insecticidal activity, *Streptomyces griseoviridis*.

Advanced rates of agricultural progress necessitate a prompt solution of disease control and growth stimulation problems in cultivars. Plant protection holds a special place in life maintenance system since, on one hand, it prevents huge harvest losses while, on the other hand, massive efforts should be mobilized to evade adverse effects caused by application of array of agents hazardous for species and the environment. Pesticides are known to produce toxic metabolites responsible for contamination of environmental media and negative impact on soil biological species. As a countermeasure, comprehensive system of plant protection based or combined action of chemical and biological agents are introduced to suppress microbial pathogens and insect pests and to control harmful and beneficial populations in agroecosystems. National programs aimed at double reduction of chemical pesticide application by compensatory supplies of biocontrol analogs have been launched in countries with elevated pesticide load (up to 19 kg/ha), like USA, Netherlands, Belgium, France /1/.

In Belarus search for biological substitutes was motivated by environmental, social and lately by economic reasons. In recent years constant rise of market pesticide prices triggered the tendency of drastic cuts in volume of imported reagents. Correspondingly, the levels of antipest and antipathogen treatment are steadily declining, affecting efficiency of plant culture.

National RLD program on industrial biotechnology conceived as a possible solution strategy envisages designing biological control agents with subsequent small-scale manufacture at Belbiopharm concern plants. It is planned that a range of bioproducts and output volume will be determined by regional market demand and agricultural specialization. This will ensure a flexible response to market fluctuations, in-time seasonal product supply and reduced transport and storage costs. Research efforts should be focused on selection of highly efficient strains of microbial antagonists and insect pathogens and further designing of derived resource – and energy-saving technologies of producing novel effective biocides.

The profile of lab of plant biological control, Institute of Microbiology, National Academy of Sciences, Belarus is in the main stream of this trend. Complex multiaspect studies aimed at isolation and characterization of microbial antagonists and insect pathogens promising for biological control of plant diseases and pests /2-4/. Screening of potential biocontrol agents takes into account criteria of antagonistic and insecticidal activity, competitive ability in soil biocenoses, technological flexibility. Priority of investigations into mechanisms of plant protective action of selected cultures, regulation of biosynthesis of antimicrobial and entomocidal metabolites as a basis for engineering more efficient biopesticides is emphasized. Performed research resulted in a collection of microbial strains-antagonists of phytopathogenic microflora, incorporating bacteria of genera *Bacillus, Pseudomonas, Streptomyces* and fungi of *Trichoderma* genus.

Domestic and foreign reports indicate a high plant protection potential for the above mentioned microbial species and good prospects for their application as active principle in such commercial biocides as Biofungus (based on *Trichoderma sp.,* Belgium), Binab-T (based on *Trichoderma harzianum* and *Trichoderma polysporum*, Sweden), Biotrack (*Trichoderma harzianum*, USA), Serenade, Kodiak (*Bacillus subtilis*, USA), Rhizoplus (*Bacillus subtilis*, Germany), Bactophyt and Phytosporin (*Bacillus subtilis*, Russia), Phytolavin (*Streptomyces griseus*, Russia), Rhizoplan (*Pseudomonas fluorescens*, Belarus, Russia) /5, 6/.

Our isolates match the best foreign analogs in antagonistic activity. Strain *Streptomyces griseoviridis* appears especially attractive due to a combination of antagonistic and insecticidal activities plus the ability to utilize a wide variety of substrates – from compounds of plant origin, like cellulose and lignin to persistent xenobiotics polluting soil. *S. griseoviridis* shows good growth in a broad range of pH (5 to 9) and temperatures (15 to 37^0C degrees centigrade) which, on one hand, solves some technological problems during fermentation and, on the other hand, ensures a high adaptive potential for the culture in natural biocenosis. All the above mentioned advantages contribute to an enhanced phytoprotective effect of *S. griseoviridis* against pathogenic microorganisms and pests (Fig. 1, Table 1). It was found that antagonism of *S. griseoviridis* towards phytopathogenic fungi was mainly determined by polyene antifungal antibiotics belonging to the pentaene group. A synthesized pigment complex is responsible for the high antibacterial activity of the culture. The obtained data laid the basis for a technology for producing biopesticide "Biactin". The process envisages a submerged culture of the actinomycete, recovery of native antibiotic solution and composting of waste mycelium with lignocellulose materials. This results in a liquid antibiotic to control plant pests and pathologies and lignocompost showing growth – promoting and phytoprotective action.

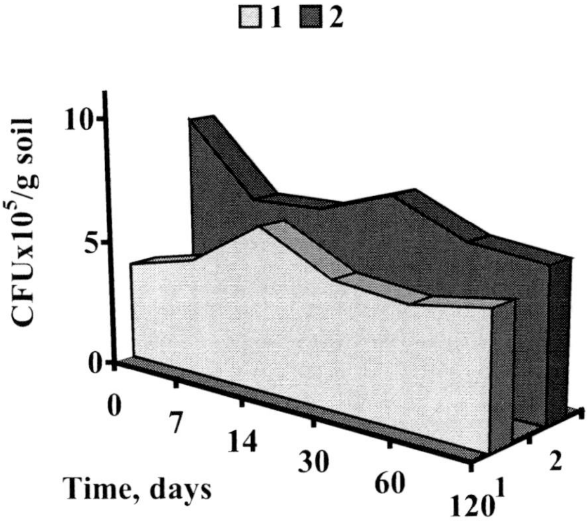

(A)

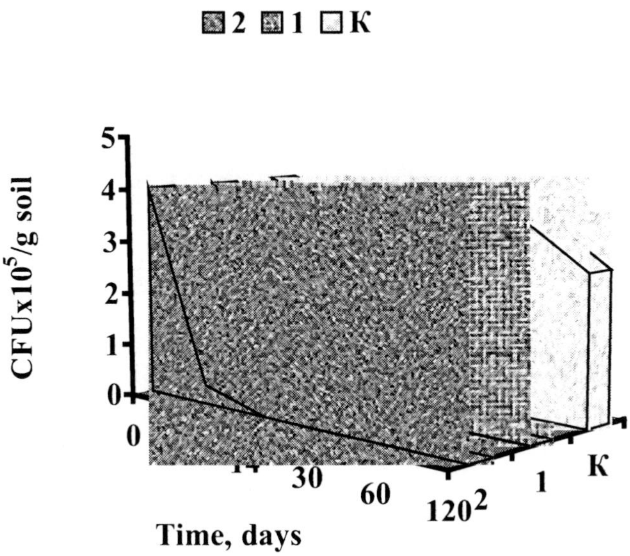

(B)

Fig. 1. Population dynamics of *S. griseoviridis* (A) and *F. oxysporum* (B) in the model system upon inoculation of actinomycete spore suspension at concentration $3,9x10^6$ (1) and $9,6x10^6$ (2) CFU /g soil. K – population dynamics of phytopathogen in antagonist-free soil

Table 1. Range of insecticidal activity of *Streptomyces griseoviridis*

Pest	Efficiency, %
Apple ermine moth (Hyponomeuta malinellus Zell.)	95-100
Apple blossom beetle (Anthonomus pomorum L.)	90-100
Spider mite (Tetranychus urticae Koch.)	85-100
Colorado potato beetle (Leptinotarsa decemlineata Say.)	75-87
Carrot leaf flea beetle (Trioza apicalis Frst.)	75-85
Tobacco thrips (Thrips tabaci Lind.)	58-75

Sporulating bacteria *Bacillus subtilis* were used in the technology of the biocide "Fruitin" designed to control canker (caused by *Nectria galligena*, *Pseudomonas syringae*), scab (caused by *Fusicladium dendriticum*), gray rot (caused by *Botrytis cinerea*) in fruit and berry cultivars (Table 2). Fruitin technology licensed for application in Belarus is now launched for industrial production. The process is based or submerged fermentation of bacteria or media with molasses – sugar-refining waste. Efficiency of the process is achieved by relatively low duration (36-42 h) and waste disposal. Biocide Fruitin is fabricated in two forms: as a liquid product (spore titer at least 8 bln per ml) and as a viscous paste (spore titer 40 bln per ml). Competitive advantage of novel biopreparation in comparison with the known analogs, like Bactophyt, Pentafag is a broad spectrum of antagonistic action – Fruitin is effective agents both fungal and bacterial pathogens, whereas Bactophyt displays mainly antifungal activity, Pentafag – antibacterial activity.

Table 2. Range of antifungal and antibacterial action displayed by strain *Bacillus subtilis BIM B-262*

Test-culture	Pathology	Diameter of growth inhibition zone, mm
	Fruit cultivars	
Nectria galligena Br.	Apple canker	35,0
Fusicladium dendriticum Str.	Apple scab	39,0
F. dendriticum X3		36,0
F. dendriticum P1		27,0
F. dendriticum V2-2		26,0
Botrytis cinerea F-15692	Berry gray rot	30,0
Pseudomonas syringae 3-3	Bacterial canker of fruit trees	37,0
P. syringae 4-1		35,0
P. syringae 2-9		38,0
P. syringae 1-1-II		40,0
	Vegetable cultivars	
Fusarium oxysporum F-55071	Fusarial rot	33,0
Alternaria brassicae	Black spot	31,0
Xanthomonas campestris 1	Bacterial wilt	33,0

The biocide application promotes rehabilitation of canker lesions in apple-trees by 46-52%, inhibits scab spread on leaves and on fruits 3- and 5-fold, respectively, and raises yields of top-grade fruits by 83% /7/. Antiscab treatment is performed during the vegetation period by spraying a 5% water suspension of liquid preparation or a 1% water suspension of paste at a gross estimate of 1-2 liters of the resultant solution per each fruit tree. To combat canker, Fruitin was applied directly on the cleaned stem lesion and then the cavity was stopped with a putty. Treatment of berry plantations was conducted in vegetation period by spraying a 2% water suspension of liquid Fruitin or 0,5% water suspension of paste to ensure both prevention and control of gray rot.

In recent years a critical situation has hit the vegetable sector of Belarussian agriculture. A spectrum of biological control agents used to control pathogens of vegetable crops is extremely limited (Trichodermin and, imported from Russia, Phytolavin and Bactophyt) whereas harvest losses induced by phytopathogenic microbial species has reached 30 to 50%. The biocide supply only partially covers the growing market demand so that farmers have to resort to chemicals. Constant application of the latter induces resistance in phytopathogens and the necessity to save crops and improve phytosanitary standards urges the use of more toxic xenobiotics.

The need to widen the scope of biological agents used to protect vegetables triggered our research efforts resulting in the technology for the ecologically safe biofungicide Phytoprotectin based on sporulating bacteria *Bacillus* intended for control of phizoctonial, fusarial pathologies, root rot, etc. (Table 3).

Table 3. Range of antimicrobial action *Bacillus subtilis 12A*

Phytopathogens	Diameter of growth inhibition zone, mm
Botrytis cinerea B1	41,0±0,5
Alternaria brassicae A1	37,5±1,0
Fusarium oxysporum P1	43,0±1,5
F. oxysporum f. sp. cyklaminis	37,0±1,0
Phytophthora cinnamomi P2	45,0±1,0
Rhizoctonia solani 2lub	39,5±0,8
Pseudomonas syringae pv. syringae 240	41,0±1,0
P. syringae pv. glycinea 280	39,5±1,0
P. cichorii 269	45,0±1,5
Xanthomonas campestris pv. campestris 33	43,0±1,5
X. beticola 8717	51,0±2,0
Erwinia caratovora pv. atroseptica 662	38,5±1,0

Laboratory and field trials have demonstrated that treatment of cabbage seeds with liquid Phytoprotectin reduced alternarial infection of seedlings 12-fold, bacterial infection by 3.5 times, phoma infection was completely suppressed. In addition to plant protection, growth-promoting effect was observed. Germination index of treated seeds increased by 47%, seedling length rose by 61 % as compared to the control. Phytoprotectin efficiency trials on cabbage seedlings are demonstrated in Fig.2.

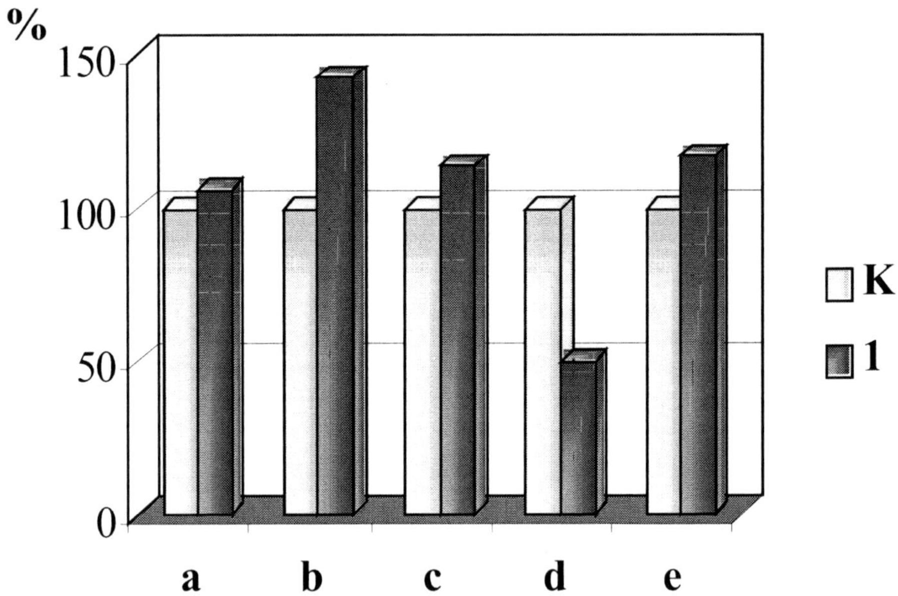

Fig.2. Effect of *Bacillus subtilis 12A* on cabbage seeds and seedlings: a - plant height, b - biomass crude weight, c - leaf surface area, d - reduction of black rot cases, e - yield of standard seedlings, %: K - control (water-steeped seeds), 1 – seeds treated with *B. subtilis 12A* cultural liquid

Elaboration of bioinsecticide Baciturin intended to combat pests of vegetable crops in greenhouse plots appears an extremely promising project of lab researchers. Baciturin active principle is made up by β-exotoxin and spore crystalline complex of bacteria *Bacillus thuringiensis*. The production process is based on submerged culture of bacteria on plentiful substrates – barley meal or molasses and fodder yeast. Baciturin is manufactured in liquid and pasty form. Following registration trials bioinsecticide Baciturin has been recommended for use against spider mite and aphides – widely distributed green house pathogens and to control Colorado beetle and carrot leaf flea in field plots. The biological efficiency of Baciturin for suppressing the Colorado beetle, the spider mite and the melon aphid reaches 85 to 90%, 79 to 100% and 60 to 80 %, respectively (Table 4). Summing up, the original biocide is not inferior to best foreign analogs and may successfully replace them at the national market when commercial Baciturin manufacture is started.

Table 4. Biological efficiency of Baciturin

Pest	**Biological efficiency, %**
Spider mite (Tetranychus urticae Koch.)	79-100
Carrot leaf flea beetle (Trioza apicalis Frst.)	81-88
Colorado potato beetle (Leptinotarsa decemlineata Say.)	85-90
Melon aphid m(Aphis gossypii Glov.)	60-80

It should be noted that possibility of conducting afore mentioned comprehensive studies has emerged due to fruitful cooperation of the Institute of Microbiology with other research institutions in Belarus (Institute of Plant Protection, Institute of Bioorganic Chemistry, Siberian division of Russian Academy of Sciences), other countries (National Chemical Laboratory, Pune, India; Institute of Horticulture and Floriculture, Poland, Centre of Ecological Studies, Polish Academy of Sciences). Research contacts facilitate exchange of expertise and personnel, mastering of up-to-date methods and techniques, realization of joint project engaging specialists of diverse profile.

The obtained findings point to the conclusion that provided highly active microbial antagonists and entomopathogens, novel substrates and advanced fermentation methods and devices, we could arrange in Belarus the profitable production of microbial preparations which will make a significant contribution to the progress of national agriculture and environmental protection.

REFERENCES

[1] R.Cook: *Ann. Rev. Phytopathol.*, 31, 53 (1993).

[2] E.I. Kolomiets,. N.A. Zdor, T.V. Romanovskaya et al.: *Bul. Pol. Acad. Sci., Biol. Sci.*, 47 (2-4), 89 (1999).

[3] T.V. Romanovskaya, E.I. Kolomiets, N.A. Zdor et al.: *Bul. Pol. Acad. Sci., Biol. Sci.*, 50 (2), 105 (2002).

[4] E.I.Kolomiets, F.A. Popov, T.V.Romanovskaya et al.: *Bul. Pol. Acad. Sci., Biol. Sci.*, 50 (2), 99 (2002).

[5] A.Stewart: *Austral. Plant Pathol.*, 30, 127 (2001).

[6] N.A.Rechcigl, J.E. Rechcigl: *Environmentally safe approaches to crop disease control.* CRC Press LLC, New York, 1997.

[7] L.N.Grigorcevich, R.V.Supranovich, E.I.Kolomiets, V.S. Dichkovskaya: *Plant protection*, 4, 40 (2002) (in Russian).

In: Biotechnology in Agriculture and the Food Industry ISBN 1-59454-119-1
Editor: G. E. Zaikov, pp. 109-114

Chapter 16

HORIZONTAL GENE TRANSFER FROM AGROBACTERIA TO PLANTS IN NATURAL AND LABORATORY CONDITIONS

T. V. Matveeva[1], N. V. Frolova[1], Yu. N. Isakov[2], O. S. Mashkina[2] and L. A. Lutova[1]

[1]St. Petersburg State University, 7/9, Universitetskaya emb., 199034, St. Petersburg, Russia, fax: +7(812)4287733,
[2]Research Institute for Forest Genetics and Breeding, 105, Lomonosov st., 394087, Voronezh, Russia, fax (0732)53-94-36
e-mail: radishlet@yahoo.com

ABSTRACT

Agrobacterium tumefaciens and *A. rhizogenes* are soil bacteria, which transfer fragments (T-DNA) from a large tumor-inducing (Ti) plasmid or root-inducing (Ri) plasmid respectively to their plant host, where these fragments are integrated into the plant genome, causing crown gall or hairy root disease. At the same time some *Nicotiana* species normally poses in their genomes sequences homologous to the T-DNA from the *Agrobacterium* as a result of horizontal gene transfer during their evolution. Some researchers consider that tumors appearing on hybrids, obtained from crosses between such species with other Nicotiana species, are induced because of activation of oncogene expression. Similar tumors appear in case of some transgenic plant lines obtained by transformation with agrobacterial oncogenes. This review is devoted to the above-mentioned above examples of horizontal gene transfer in natural and lab conditions.

Key words: horizontal gene transfer, *Nicotiana*, *Raphanus sativus*, tumorigenesis

INTRODUCTION

Transfer of genetic material, existing among organisms, which could not be termed as parents and progeny, is called: "horizontal gene transfer". For the procaryotic organisms it is a typical mechanism of appearing of combinative variability. Existence of horizontal gene transfer in the evolution of eucaryotic organisms was called in question for a long time. The first fact supporting this theory was the discovery in 70th of natural vector system, based on soil bacteria of the genus *Agrobacterium*. These bacteria transfer fragments (T-DNA) from a large tumor-inducing (Ti) plasmid or root-inducing (Ri) plasmid respectively to their plant host. Once inside the cell, these fragment traffic to the plant cell nucleus where they are integrated into the plant genome [1-4]. Laboratory experiments show, that it is possible to regenerate whole plants from the transgenic tissues. On the other hand, there are data, pointed that a number of "untransformed" plants, such as some *Nicotiana* species, already contain sequences, homologous to the T-DNA from *Agrobacterium rhizogenes*, in their genomes, i.e. horizontal gene transfer (HGT) took place in the evolution of genus *Nicotiana*. This mini review is devoted to the description of the examples of HGT in *Nicotiana* species. Besides, we discuss the possibility of usage of horizontal transfer of *Agrobacterium* oncogenes into other plant genomes (in the lab conditions) for obtaining of morphological abnormalities similar to ones, described in Nicotiana hybrids, where T-DNA-like sequences are present.

T-DNA-LIKE SEQUENCES IN *NICOTIANA* GENOMES

In 1982 White *et al* found T-DNA-like sequence in genome of untransformed *Nicotiana glauca*. They discovered the DNA sequence that showed more than 80% homology with *A. rhizogenes* T-DNA, designated c-DNA [5]. Southern hybridization indicated that the cT-DNA contained the sequences of ORFs 11 *(rolB),* 12 *(rolC),* 13, 14, and 15 *(rolD),* which were denoted *NgrolB, NgrolC, NgORF13, NgORF14,* and *NgrolD,* respectively. The cT-DNA was represented by an imperfect inverted repeat, which carried a single copy of the *NgrolB* gene and two copies of the remaining genes [6]. In the *N. tabacum* genome, the genes homologous to *A. rhizogenes rol* genes were also found. They were denoted as *trol* genes or *tORF* [7, 8]. A hypothesis has been advanced that these genes originated by the horizontal gene transfer during the evolution of the *Nicotlana* genus. Later homologs of A.rhizogenes oncogenes were found in other Nicotiana species (table 1) [9]. RolB homologs were found in 3 additional to known species ORF13 and ORF14 homologs were found in 4 additional species. The biggest list contained Nicotiana species with rolC homolog (13 species, including *N.glauca* and *N.tabacum*). Authors discuss this phenomena in terms of cT-DNA rearrangements during the *Nicotiana* evolution. But it can be also explained as higher conservation of rolC sequence, comparing to other oncogenes, and as a result its easyer detection.

Table 1. Presence of Introgressed Genes in Nicotiana species (Intrieri, Buiatti, 2001)

Species	Gene			
	rolC	rolB	ORF 13	ORF 14
N. *glauca*	+	+	+	+
N. cordifotia	+	+	+	+
N. benavidesii	+			
N. tomentosiformis	+		+	+
N. otophora	+		+	+
N. setchelli	+			
N. tabacum	+		+	
N. arentsii	+			
N. acuminata	+			
N. miersi		+		
N. bigelovii		+		
N. debneyi	+			
N. gossei	+			
N. suaveolens	+			
N. exigua	+			

ORIGIN OF CT-DNA

The presence of sequences homologous to agrobacterial oncogenes in the genomes of several tobacco species supports the hypothesis on the existence of horizontal gene transfer (HGT) between the representatives of the two suprakingdoms: prokaryotes and eukaryotes. However, the direction of HGT was the subject to debate for a long time [10, 11]. Only in 2002 Suzuki et al. [12] clarified the situation. More complete sequence of T-DNA insert was obtained by these authors for Nicotiana tabaccum, N. glauca, N.tomentosa, N.tomentosiformis. They found that a sequence homologous to cT-DNA extended outside NgORF14 in the mikimopine-type pRi1724 T-DNA. Besides, authors identified plant boundary DNA sequences. Basing on the difference in these sequences authors concluded, that oncogenes were transferred to Nicotiana from Agrobacterium, and that there were several independent transformation events. Besides, authors proposed that T-DNA from some other Ri plasmid was integrated in the genomes of N.benavidesii and N. otophora.

EXPRESSION OF T-DNA-LIKE GENES IN DIFFERENT NICOTIANA SPECIES

Analysis of expression of these cT-DNA genes demonstrated that it was tissue-specific and occurred both in the parental species and tumorous hybrids. Thus, the *tORF13* was expressed in sepals, petals, apexes, and young leaves; the *trolC* gene was expressed in young tissues of *N. tabacum* [7, 8]. Expression of the *NgrolB* and *NgrolC* genes was detected in *N.*

glauca stems [10]; the *NgrolB, NgrolC, NgORF13,* and *NgORF14* genes were expressed in genetic tumors formed on *N. giauca* x *N, langsdorffii* hybrids and not in their leaves [10, 13-14]. Nagata *et al.* [15] studied expression of the *NgrolB* and *NgrolC* genes in genetic tumors initiated on the cut-off tobacco leaves and showed that the *NgrolB* gene was expressed in all proliferating cells in the region of the cut beginning from the third day after explanation, while the *NgrolC* gene was expressed in the developed tissues, such as pro-cambium and teratomas. During the regeneration of tumorous tissues into plants, a gradual decrease in the expression of these genes was observed. The authors suggested that the *Ngrol* genes were involved in the genetic control of tumor formation: the *NgrolB* at the mitosis stage, and the *NgrolC* at the stage of tissue differentiation [15].

Intrieri and Buiatti investigated cT-DNA genes expression in all species in which the corresponding DNA sequences had been found by them. Gene expression was differentially regulated in different species. Particularly, *rolB* mRNA was found to be present in all cases in hormone autotrophic (habituated) callus tissues, but not in leaf expl ants. *RolC* was expressed in both callus and leaves of *Rustica* species, *N. glauca* and *N. cordifolia,* and only in callus in species belonging to the subgenus Petunioides; no expression at all was shown to occur in species of the subgenus Tabacum. ORF13 and 14 were always expressed in callus ORF13, being transcribed also in leaves of *N. tabacum* and *N. tomentosiformis* [9].

Some authors discovered hormonal regulation of the *Ngrol* genes. The expression level of the *NgrolB* was positively regulated by auxin. The reverse picture was observed with the *NgrolC:* This gene was negatively regulated by auxin [13]. In cultured *N. tabacum* leaf tissues, the accumulation of *trolC* mRNA was negatively regulated by auxin and induced by cytokinin [7]. The regulatory regions of the *Ngrol* genes are simitar to those of the Ri *rol* genes. In the pairs *NgrolB-Ri rolB, NgrolC-Ri rolC*, the same regulation mechanism by auxin (NAC) and similar tissue-specificity of expression were shown [15].

Effect of cT-DNA Expression on Phenotype of Plants

In 1930 Kostoff described tumor formation in Nicotiana hybrids. Naf subdivided all tobacco species into two groups: + and -. Basing on their morphological characteristics, species of + group was also named shoot-prodycing, species from – group named root-producing. Crosses within one group do not lead to tumor formation. F1 hybrids obtained from crosses between group + and group -species, derived amphidiploids, tri-, tetra-, and pentaploids with different ratios of parental genomes are tumorous [17].

The distribution of *rol* genes in the genus Nicotiana reported by Intrieri and Buiatti [9], showing their absence in all shoot-producing species, and presence in rooting-prone *Nicotiana.* Besides these authors shown that rolB is expressed only in hormone-anitotrophic tumorous tissue (habituated callus). Pattern of expression of other genes is different, but all of them are expressed in tumorous tissues [9-10, 13-15].

Basing on these data it is possible to conclude, that *Nicotiana* hybrids in which one of the parental species is shoot prone and therefore lacks the *rol* complex, and the second one carries the *A. rhizogenes* genes, will form spontaneous tumors due to the combination of two factors: *rol*-genes expression and "shooty genes" expression [9]. However, the finding that cT-DNA genes were expressed not only in the tumorous hybrid but in the parental species as well casts some doubt on their participation in tumor formation in tobacco. Furthermore, not all of these genes code for functional products [10,16]. Thus, the *NgrolB* gene carries two point mutations affecting its functions, while the activity of the *NgrolC* gene is preserved from the event of *Agrobacterium*-mediated transformation that occurred in the ancestral tobacco species [10].

Tumorigenesis as a Result of Presence of Agrobacterial Oncogenes in Plant Genomes

Another facts supporting the theory about role of *Agrobacterium rhizogenes* oncogenes in *Nicotiana* tumor formation is tumorous phenotype of transformed with agrobacterium oncogenes plants from other taxons.

In most cases transformation of plants with *Agrobacterium* oncogenes leads to formation of transgenic tissues: tumors in case of *A. tumefaciens* or hairy roots in case of *A. rhizogenes* [1-4]. However, in some cases atypical response to agrotransformation was observed. Some plant genotypes after transformation with *A. rhizogenes* strains are able to form tumors, similar to tumors of *Nicotiana* hybrids [4, 18].

Transgenic plants, regenerated from transgenic tissues with agrobacterial oncogenes are characterized by multiple morphological abnormalities [19]. Reseachers were able to obtain transgenic plants with whole set of *A.rhizogenes,* but not *A.tumefaciens* oncogenes. There were not tumor forms among these transformants. At the same time it is possible to obtain transgenic plants carrying single *A.tumefaciens* oncogenes. They also characterized by modified phenotype. There are forms with tumors among plants transformed with single *tmr* or *tml* locus (but not *tms*) from *A.tumefaciens* [20]. One of the examples of such experiments (which is done in our laboratory) is transformation of normal plant of radish *Raphanus sativus* with construct, containing *tmr* gene under CaMV 35S promoter [21]. Lines with this construct form tumors on the crop roots similar to those from radish lines with genetic tumors. It is a suitable model for investigation of radish tumor formation. Another possible model for such experiments is *Betula pendula*. This species has two morphological variants: normal and Karelian birch. Clones of karelian birch have a lot of similarities to radish tumor lines (including changes of morphology *in vivo*, secondary tumor-like structures *in vitro*) [22-24]. We can speculate that transformation of normal birch with *tmr* gene will lead to formation of a phenocopy of karelian birch.

Conclusion

By now, examples of horizontal gene transfer during the evolution of plant species, was clearly demonstrated in case of only one Genera (*Nicotiana*). Search for new examples of HGT is extremely important, because the discovery that bacterial genes are naturally present in food crops may help to assuage public concerns about the use of genetically engineered foods. This is especially timely now when the genetic engineering of food crops is being attacked as an unnatural laboratory process which should be banned. If somebody demonstrates that T-DNA from either *A. tumefaciens* or *A. rhizogenes* is present in other plants, including food crops, then people have been eating naturally genetically engineered food for centuries.

Using of HGT from *Agrobacteria* to plants in the laboratory conditions could also be important for fundamental and applied science. Such plants may have traits valuable for the national economy.

ACKNOWLEDGEMENTS

The research described in this publication was made possible in part by Award No. ST-012-0 of the U.S. Civilian Research and Developmental Foundation for the Independent States of the Former Soviet Union (CRDF) and Award No 020448800 of RFBR.

REFERENCES

[1] J. Schell, M. Van Montagu, M De Beuckeleer, et al.: *Proc R Soc Lond B Biol Sci*, 204(1155),251-66 (1979).
[2] J. Schell, M. Van Montagu, *Basic Life Sci*; 9:159-79(1977).
[3] F.F. White, B. H. Taylor, G. A. Huffman, et al. *J. Bacteriol,* 164(l), 33-44.(1985).
[4] F. F. White, G. Ghidossi, M. P. Gordon, and E. W. Nester, *Proc.Natl.Acad. Sci. USA*, 79,3193-319 (1982).
[5] F.F. White, D.J. Garfinkel, G.A Huffman. et al., *Nature*, 301 (5898), 348-350. (1983)
[6] I.J. Furner, G.A. Huffman, R.M. Amasino, et al., *Nature*, 319, 422-427 (1986).
[7] A.D. Meyer, T. Ichikawa, F. Meins, *Mol. Gen. Genet*., 249, 265-273. (1995)
[8] C. Frundt, A.D. Meyer, T. Ichikawa, F.J. Meins. *Mol. Gen. Genet.,* 259(6), 559-568. (1998)
[9] M.C. Intrieri, M. Buiatti. *Molecular Phylogenetics and evolution*. 20(1), 100-110 (2001)
[10] S. Aoki, K. Syono, *Proc. Natl. Acad. Sci. USA*, 96(23), 13229-13234 (1999)
[11] F. Bertolla. P. Simonet, P., *Res. Microbioi*, 150(6), 375-384 (1999)
[12] K. Suzuki, I. Yamashita and N.Tanaka *Plant J*., 32(5), 775-787 (2002)
[13] T. Ichikawa, Y Ozeki, K.Syono, *Mol, Gen. Genet*., 220(2), 177-180. (1990)
[14] S.Aoki, A. Kawaoka, M.Sekine, et al., *Mol. Gen. Genet*., 243, 706-710. (1994)
[15] N. Nagata, S. Kosono, M. Seldne, et al. *Plant Cell Physiol,* 36(6), 1003-1012. (1995)
[16] K. Lemcke, T. Schmutiing, *Plant Mol. Biol*., 36 (5), 803-808. (1998)
[17] T. Ichikawa K. Syono. *Plant Cell.Physiol*. 1991 32(8) 1123-1128
[18] L.A. Lutova, Z.B. Pavlova, M.M. Ivanova *Russian J. Of Genet*. 34(2), 165-182 (1998)
[19] Tepfer D. *Cell* 37, 959-967 (1984)
[20] L. A. Lutova N.A. Provorov, O.N. Tikhodeev et al. *Genetics of plant development,* Nauka, St. Petersburg., 2002 539 p. (in Russian)
[21] T.V. Matveeva, I.S. Buzovkina, N.V. Frolova, et al. *International Symposium "Biotechnology approaches for exploitation and preservation on plant resources"*. Yalta. Ukraine. p.68. (2002)
[22] Narbut S.I. *The bulletin of the Leningrad university*, 15, 144-149. (1967) In Russian
[23] I.S. Buzovkina, I. Kneshke, L.A.Lutova *Russian J. Of Genet*. 29(6), 1002-100 (1993)
[24] .S. Mashkina, T.M. Tabatskaya, I. Yu. Isakov. Forest sector development problems / *Extended Abstracts. Petrozavodsk*. 1998, 10 - 12. (1998)

In: Biotechnology in Agriculture and the Food Industry ISBN 1-59454-119-1
Editor: G. E. Zaikov, pp. 115-124

Chapter 17

ORGANIC FERTILIZERS FROM WOOD WASTE PRODUCTS, RECEIVED WITH THE HELP OF BIODESTRUCTOR BARKON

O. V. Sviridova, N. I. Vorobyov, V. B. Petrov and V. V. Kochetkov
The State All-Russian Research Institute for Agricultural Microbiology
Podbelski sh., 3, Saint-Petersburg, Pushkin-8, 196608, Russia,
Fax: (812) 470-51-00, e-mail: osviridova@newmail.ru

ABSTRACT

Biotechnological recycling of wood waste products into the components of humus substances and their application as an organic fertilizer can increase the fertility and preservation of humus fund in soils. For the specified purposes the technology of manufacturing special biological preparation BARKON – biodestructor of wood waste products of coniferous trees has been developed. Biodestructor BARKON contains a lignin destructing complex of fungi (*Penicillium, Trichoderma, Aspergillus*) and bacteria. The organic fertilizer obtained with use of biodestructor BARKON does not contain phytopathogenic fungi and suppresses development in soil.

Key words: biodestructor, microorganisms, bark, lignin, coniferous trees, maximal coefficients of correlation.

Products of the decomposition of wood and bark (lignin, cellulose) can be one of the basic components participating in formation of humus in soil [1, 2]. Wood of deciduous trees contains 20-30 % lignin, and coniferous trees - up to 50 %, the contents of cellulose in a trunk of wood - up to 40-55 %. Under natural conditions the bark of trees, especially coniferous, decays tens years and can be toxic [3, 4]. In this relation, the research problem is to create the complex of microorganisms and to chouse of optimum conditions for acceleration of destruction wood waste products of coniferous trees into humus like forms.

OBJECTS AND METHODS

Object of research was functionally stable complex of microorganisms which is capable to decompose effectively a bark and wood of coniferous trees to organic fertilizer. Structure of a complex formed at growing of the microorganisms isolated from natural microbiocenosis on special selective media (Certificate "A Mode of Decomposition of Wood", Sviridova O.V. et al., 1993) [5].

NPK dozes for composting a bark and sawdust were selected on the basis of a recommended ratio of organic carbon to the total nitrogen (25:1). Variants of experiment differed in titers of microorganisms and dozes of nitrogen (C/N 25:1 and 50:1). At the end of experiment (in 90 day - for experiment with a bark and 20 day - for sawdust) we determined number of the basic trophic groups of microorganisms (proteolytic, humus-decaying, celluloselytic, amylolytic bacteria and micromycetes) [6] and C, N, N-NO_3, CO_2 contents [7]. The degree of destruction of a bark and sawdust of coniferous trees was estimated visually (on 1- 10 points of a scale) and on action on seedlings of a garden radish (soaked in solutions of composts).

For definition of similarity of treatment, the maximal coefficients of correlation between parameters, streams of biotransformation and C/N ratio in a biomass of microorganisms were calculated. The resulted dendrograms and graphs were constructed by using correlation, clusters analyses and analysis of variance.

RESULTS AND DISCUSSION

Date of experiments shown, that after 90 days of composting the bark for control variant was the beginning of destruction (level of destruction – 1 point), that explained the stability of bark of coniferous under natural conditions (Table 1) [8].

Date on composting the bark and sawdust have shown, that in variants with more active decomposition substantial increase in number was marked practically for all groups of microorganisms. The increase in number of proteolytic and humus-decaying bacteria in these variants revealed destruction of cyclic chains of lignin – base component of bark and wood of coniferous trees as these microorganisms produce peroxides [9].

Cluster analysis allowed us to choose the microorganisms, which have maximal coefficient of correlation with level of destruction for bark and constructed graph of maximal correlations (Fig.1). The scheme demonstrated a strong positive correlation ($r=0, 78\pm0, 05$) between intensity of decomposition and number of proteolytic bacteria, humus-decaying bacteria and "white rot" fungi was revealed. There fore, these bacteria produce peroxides to oxidize organic substrate and to destruct the aromatic chains. During the destruction of organic substrate simple compounds formed, which are used the other group of microorganisms, consisting of celluloselytic, amylolytic and nitrogen-fixing bacteria. Thus, graph determined basic complex of microorganisms, required for decomposition of bark. The complex consists of proteolytic, humus-decaying, celluloselytic, amylolytic and nitrogen-fixing bacteria and "white rot" fungi. The periphery of graph consists of micromycetes, actinomycetes and nitrificators, which can utilize the metabolic products of basic complex. Number of micromycetes has negative correlation with amount of different nitrogen forms. It was shown that micromycetes use different nitrogen forms, it is suggested possibility of formation of two chains of interaction of microorganisms (carbon- and nitrogen – chains).

It was noted that for an active biological decomposition of bark of coniferous trees the presence of bacteria (*Pseudomonas*, *Mycobacterium*, *Bacillus*, *Cellulomonas, etc.*), micromycetes (*Penicillium*, *Aspergillus, Trichoderma,*) and "white rot" fungi is required.

The study of sawdust destruction with application of selected lignin destructing complex shown tits high efficiency (15-20 days at composting of sawdust to organic fertilizer) (Table 2). For an active biological decomposition of sawdust of coniferous trees the presence of bacteria (*Pseudomonas*, *Mycobacterium*, *Bacillus*, *Flavobacteriu*, *etc.*) and micromycetes (*Penicillium*, *Trichoderma*) is required.

Graphs of maximal correlations (Fig.2) specified the role of each group of microorganisms in biotransformation of wood wastes. The scheme had shown that strong positive correlation groups of proteolytic, humus-decaying and amylolytic bacteria with level of substrate destruction and negative correlation with the content of carbon. It was noted an alternative in the scheme of bark destruction, that for sawdust destruction increase the role of amylolytic bacteria in basic complex. This fact can explained by the present of cellulose on the sawdust and by rapid formation of simple compounds. This scheme show that micromycetes has negative correlation with destruction of sawdust that can inhibit processes of decomposition of sawdust.

Dendrograms shown that level of similarity between variants on composting bark and sawdust depends on the conditions of cultivation of the microbial community (Figs. 3, 4). Different complexes of microorganisms were allocated on the selective media. Decrease of nitrogen dozes practically don't change effect of complex BP2 and this variants (1/2 NPK +BP2 or 1/2 BP2) were clustered together. Clusters analysis allowed us to choose the effective complex for composting – BP2, containing basic complex of microorganisms.

The streams of biotransformation in experiment on composting bark and sawdust was demonstrated on the graphs (Figs. 5, 6). These graphs shown the role of each group of microorganisms in biotransformation of nitrogen and carbon. The positive correlation of proteolytic bacteria with the content of nitrates revealed that they participated in the formation of mineral nitrogen. Proteolytic bacteria used organic nitrogen and produced ammonium, with nitrificated by the other groups of microorganisms which form the nitrates. The role of proteolytic bacteria as microorganisms starting trophic circuit is visible, but the use of only these bacteria for destruction of wood waste products can decreased the composting quality. Also functioning of only micromycetes (Fig. 6) can result in suppression of plants growth.

The checking quality composts from variants with a strong and average degrees of destruction has shown stimulation of emergency of garden radish, i.e. absence of toxicity.

Besides the inhibiting effect of composts on development of a phytopathogenic fungus *Fusarium oxysporum (Shlecht.) Snyder et Hanssen* was revealed. The organic fertilizers obtained with use of biodestructor BARKON do not contain phytopathogenic fungi and suppress their development in soil.

Table 1. Data on composting bark of coniferous trees.

C_{org}/N_{total}	Parameters	NPK 1	NPK+ BP1 2	NPK+ BP2 3	NPK+ BP3 4	NPK+ BP4 5	NPK+ BP5 6	NPK+ BP6 7
30.83	Micromycetes	3.5	1.7	1	1.6	1	3.4	1.5
33.12	Proteolytic bacteria	226.7	221.3	319.3	308	240	156	139.3
32.71	Amilolytic bacteria	225.3	264	210.3	220	176	359.3	436.7
33.07	Humusdecaying bacteria	114.9	12.8	108.8	126.3	75.1	101.3	89.6
33.07	Celluloselytic bacteria	1.5	1.2	9.5	0.8	2.5	1.5	0.2
	N_{total}	1.95	1.9	2.2	2.32	2	12.2	2.1
	$N\text{-}NO_3$	9.2	84.3	12	18.9	107	11.2	74.9
	Destruction	7	8	10	10	8	9	9
	$C\text{-}CO_2$	2.34	2.1	1.9	1.8	1.7	2.3	1.8
	Mass roots of radish	82	85	102	101	97	106	86

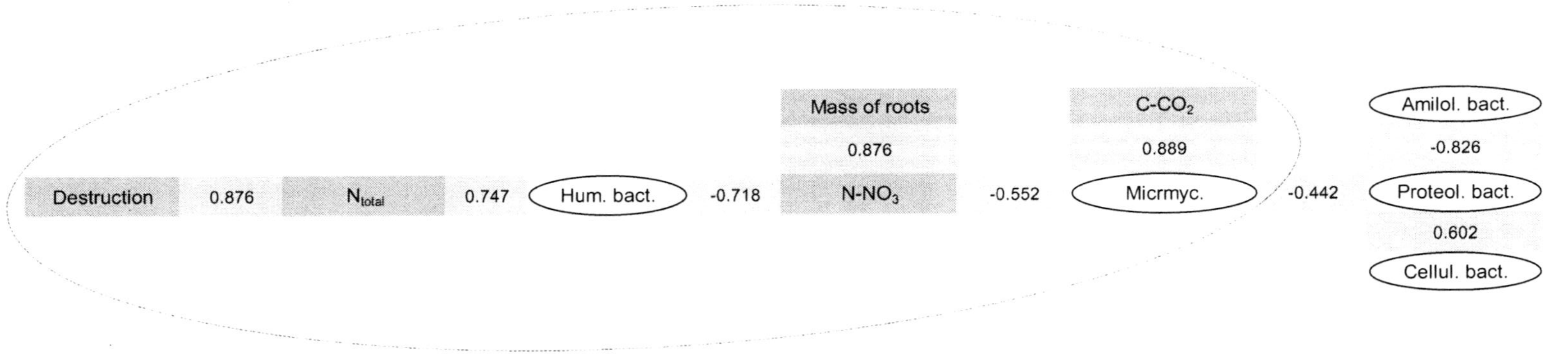

Fig.1. Graph of maximal correlations with parameters on composting bark of coniferous trees

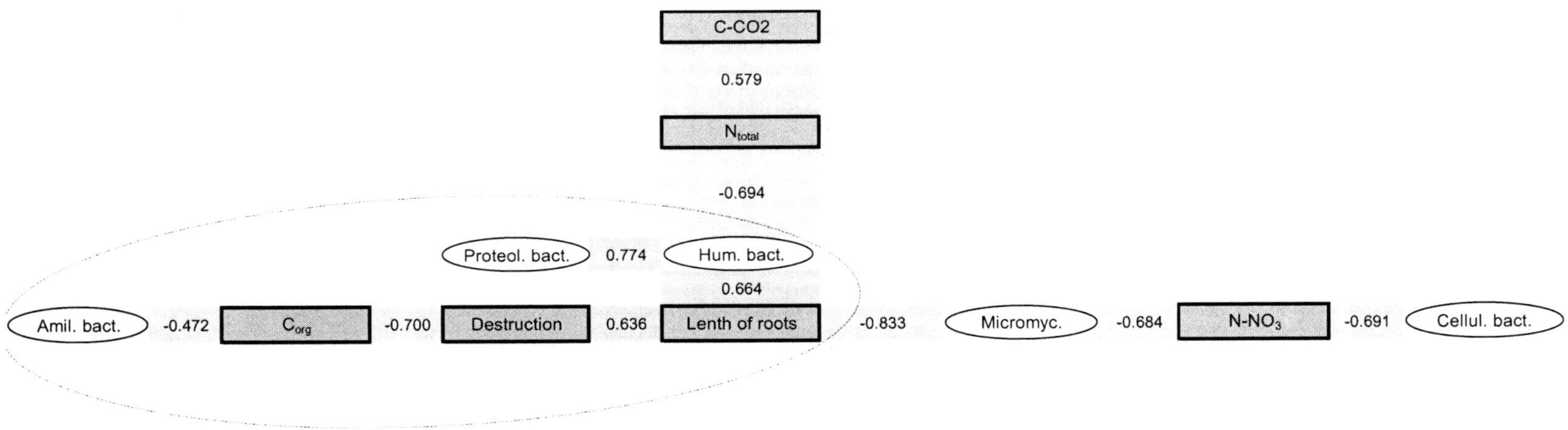

Fig. 2. Graph of maximal correlations with parameters on composting sawdust of coniferous trees.

Table 2. Data on composting sawdust of coniferous trees.

C_{org}/N_{total}	Parameters	NPK 1	1/2NPK 2	NPK+ BP1 3	NPK+ BP2 4	NPK+ 1/2BP2 5	1/2NPK+ BP2 6	1/2NPK+ 1/2BP2 7
30.83	Micromycetes	129.7	76.7	87.3	21	18	28	37.3
33.12	Proteolytic bacteria	2260	266.7	500.	600	500	833	4466
32.71	Amilolytic bacteria	110	26.7	123	20	170	50	86.7
33.07	Humus decaying bacteria	3.7	3.3	4	4	6	12.7	21
33.07	Celluloselytic bacteria	450	45	2.5	95	45	45	7.5
	N_{total}	1.84	1.02	1.82	1.69	1.59	1.05	0.91
	$N\text{-}NO_3$	1.7	4.4	2.9	4.3	4	3	3.5
	C_{org}	42	43.5	42.2	42.09	42.16	42.34	42.15
	Destruction	5	3	9	10	7	7	8
	$C\text{-}CO_2$	0.29	0.35	0.51	0.45	0.33	0.21	0.29
	Radish mass	3.41	3.18	2.9	5.23	2.92	4.31	3.05
	Lenth roots of radish	29	29.3	34.6	40	38.3	45	41.6
	NPK	1	0.5	1	1	1	0.5	0.5
	BP	0	0	1	1	0.5	1	0.5
	C_{org}/N_{total}	24.71	9.89	14.55	9.79	10.54	14.11	12.04

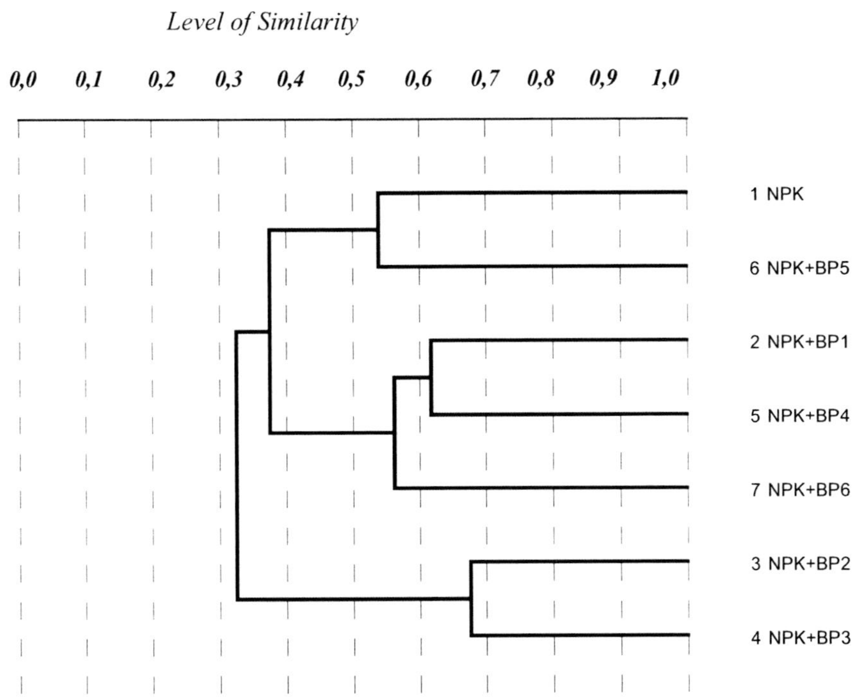

Fig. 3. Level of similarity between variants on composting bark of coniferous trees.

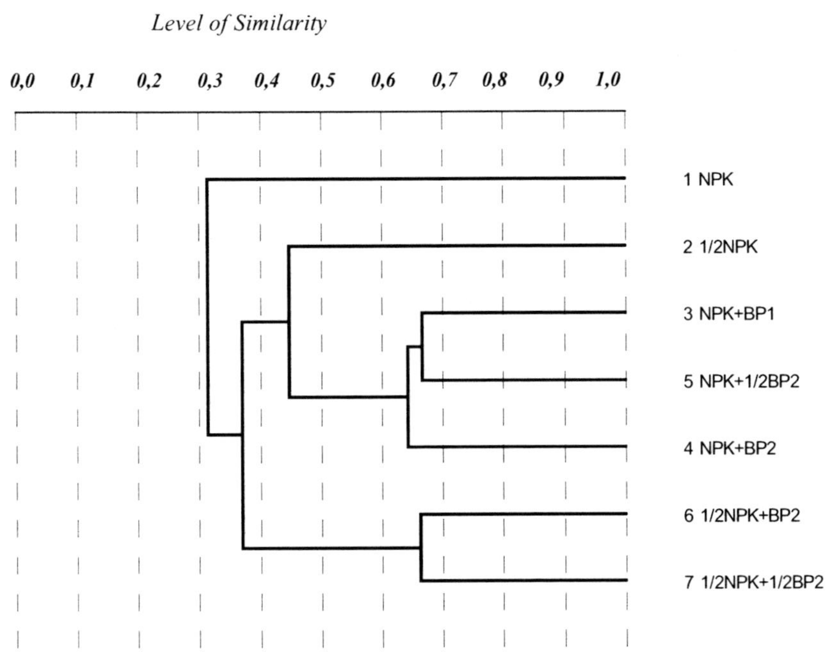

Fig. 4. Level of similarity between variants on composting sawdust of coniferous trees.

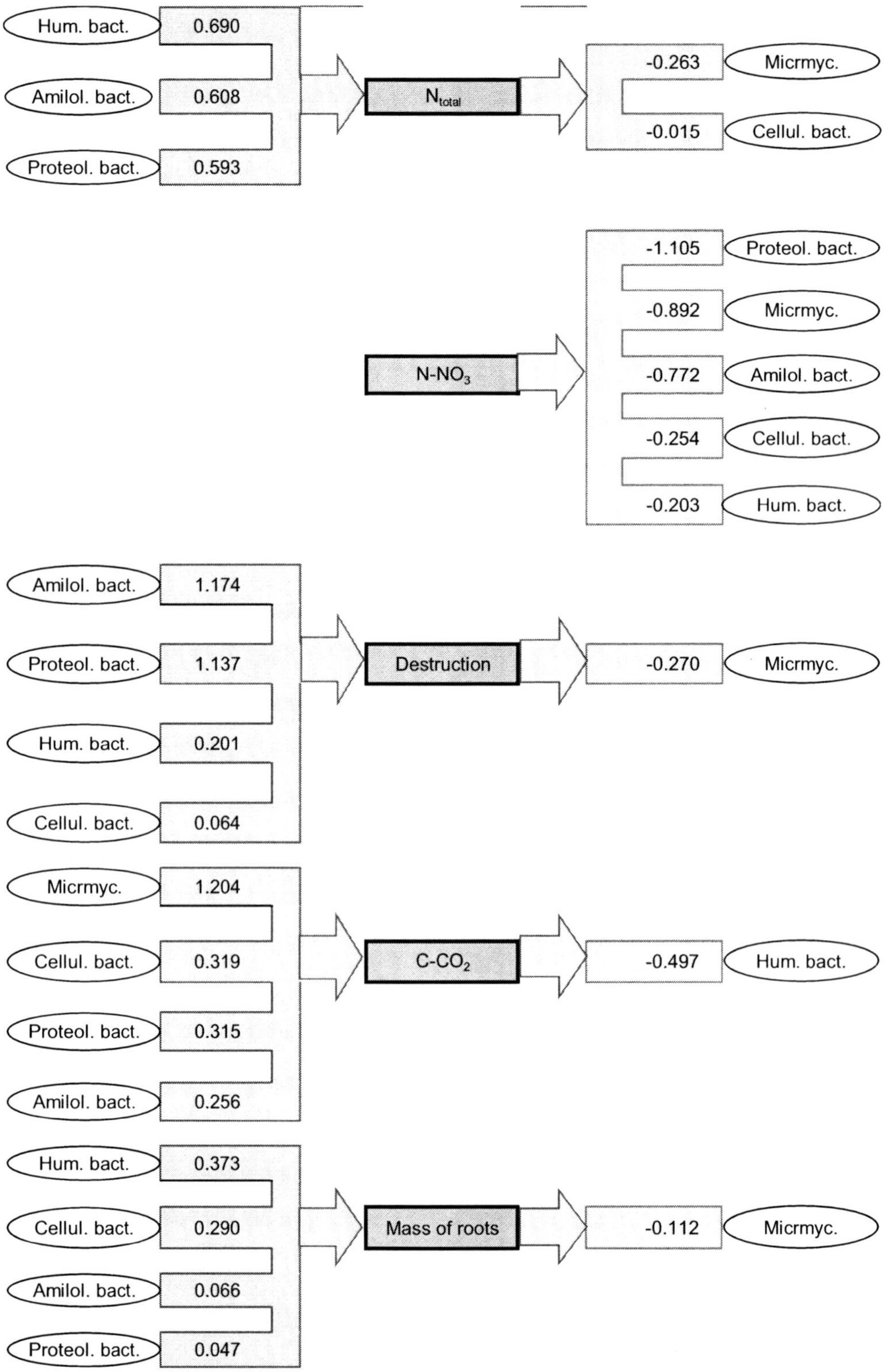

Fig. 5. Graph of streams of biotransformation in experiment on composting bark of coniferous trees.

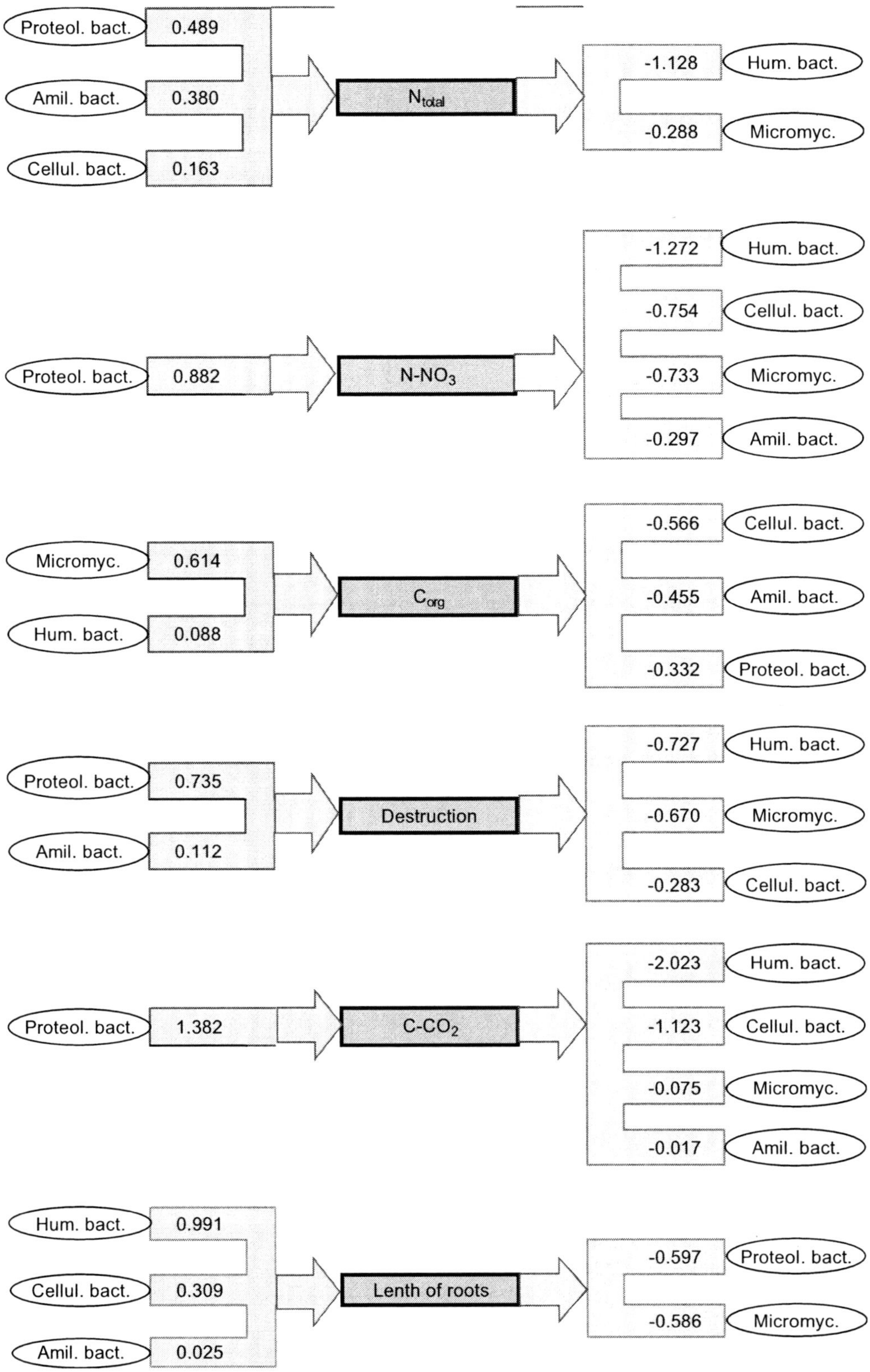

Fig. 6. Graph of streams of biotransformation in experiment on composting sawdust of coniferous trees.

Application on the soil of wood biocomposts stimulated the growth of a garden radish and winter rye (early growth, formation of an elevated and root biomass). The increase of biomass winter rye after of use biocomposts from sawdust of coniferous trees was 8-15 % over control. Optimum dozes of application biocomposts wood waste products were 3 tons per hectare.

Thus, the application organic fertilizers from wood waste products can increase fertility and preservation humus in soils.

Conclusion

(1) Stable complex of bacteria and micromycetes for decomposition of bark and sawdust of coniferous trees was formed.
(2) Optimal conditions for function of this stable complex for composting the bark and sawdust of coniferous trees were selected.
(3) Application of this complex of microorganisms allowed us to receive composts for increasing fertility and sanitation of soil.

References

[1] M.M. Kononova: *Organic substance of virgin and mastered soils. Science, Moscow*, 1972, p.-p. 7 - 69. (in Russian).
[2] D.S.Orlov: *Humic acids of soils*. MGU: Moscow, 1974, 382 p. (in Russian).
[3] E.V. Melnikov, V.N. Petry: *Wood J.,* № 2, Archangelsk, 1963, p.-p. 120-125. (in Russian).
[4] N. Kokalisburelle, R. Rodrigezkabana: *Plant and Soil*, 162 (2), 1994, p-p. 169 - 175.
[5] O.V.Sviridova et al.: "*A Mode of Decomposition of Wood*". Certificate № 1792974, 1993. (in Russian).
[6] N.S.Egorov: *Practicum for microbiology*. MGU: Moscow, 1983, 215 p. (in Russian).
[7] E.V. Arinushkina: *Guidance for chemical analysis of soils*. MGU: Moscow, 1970, 473 p. (in Russian).
[8] O.V.Sviridova, L.V.Mikhaleva, N.I.Vorobyov, V.V. Kochetkov: *Mycology and phytopathology J.*, 35 (6), 2001, p.p. 38-47. (in Russian).
[9] E.M. Mishustin, V.T.Emtcev: *Microbiology*. Moscow, 1987, 357 p. (in Russian).

In: Biotechnology in Agriculture and the Food Industry ISBN 1-59454-119-1
Editor: G. E. Zaikov, pp. 125-140

Chapter 18

RHIZOBIAL NODULATION FACTORS — A NEW CLASS OF MORPHOGENIC MOLECULES WITH POTENTIAL FOR PRACTICAL USE

A. O. Ovtsyna[1], E. A. Dolgikh[1], V. I. Torgov[2], A. I. Zinin[2], J. V. Cullimore[3], J.-J. Bono[3] and I. A. Tikhonovich[1]

[1]All Russia Research Institute for Agricultural Microbiology, 196608, Saint-Petersburg, Pushkin, Podbelsky sh.3;

[2]Institute of Organic Chemistry RAS, 117913, Moscow, Leninsky pr., 47;

[3]Laboratoire de Biologie Moleculaire des Relations Plantes-Microorganismes, UMR INRA-CNRS 215, BP 27, 31326, Castanet-Tolosan, France.

e-mail: ovtsyna@yahoo.com

Fax: 7(812) 470 43 62

ABSTRACT

Nodulation factors (Nod factors) are signal molecules of soil bacteria rhizobia, which are produced during symbiotic interaction between rhizobia and legume plants. Nod factors trigger a set of plant reactions resulting in a formation of nitrogen fixing root nodules. They are active in nanomolar concentrations, possess mitogenic and morphogenic activity and determine host specificity of symbiotic interaction. In order to study Nod factor effect on plants, to reveal their receptors in plants and to analyze signal transduction induced by these molecules, the set of Nod factor analogs differing in their structure and carrying new type of fluorescent label, was chemically synthesized. Biological activity of these compounds was analyzed, possibilities of their applications are discussed.

Keywords: Nitrogen – fixing symbiosis, plant defense, rhizobia, Nod factors, plant growth regulators, signal transduction.

INTRODUCTION

Nod factors and Their Functions in *Rhizobium* – Legume Symbiosis

Soil bacteria, collectively referred to as rhizobia, are able to establish symbiosis with leguminous plants, resulting in a formation of the new organ, nitrogen-fixing nodule, on the plant root. Rhizobium- legume symbiosis is host-specific, e.g. certain plant species are able to form nodules only in interaction with certain rhizobial species. Plants forming nodules with the same rhizobial species belong to the same cross-inoculation group (Table 1). Recognition between plant and its potential symbiont occurs via chemical signal exchange (reviewed by Spaink, 1996; Schultze and Kondorosi, 1998; Perret et al., 2000; Hirsh et al., 2001). At the first stage of this molecular dialogue plant secretes specific flavonoids, which induce expression of a set of bacterial genes (*nod*, *noe*, *nol* genes) responsible for synthesis of rhizobial signal molecule, Nod (nodulation) factor. Nod factors are the key morphogenic signals triggering a number of plant responses (membrane depolarization, ion fluxes, root hair deformations, cortical cell divisions, plant gene expression) leading to formation of root nodule (Spaink, 2000; Cullimore et al., 2001; Cardenas et al., 2000; Oldroyd, 2001). Nod factors are lipochitooligosaccharides (LCOs) - oligomers of N-acetylglucosamine (3-6 GlcNAc units) carrying fatty acid chain at non-reducing sugar and various modifications on chitin backbone (Figure 1). The structure of Nod factors is major determinant of symbiotic host specificity (Denarie et al., 1996; Schultze and Kondorosi, 1998; Spaink, 2000; Perret et al., 2000). For example, *Sinorhizobium meliloti* strains, symbionts of alfalfa, carry C16 fatty acid with two double bonds and sulfate group at the reducing terminus (Lerouge et al., 1990), whereas *Rhizobium leguminosarum* bv. *viciae* strains infecting pea produce Nod factors with C18:4 fatty acid and no modification at the reducing GlcNAc (Spaink et al., 1991) (Figure 1, Table 1). Mutants of *S.meliloti* with deleted *nodS* gene (responsible for sulfate attachment to the chitin backbone of Nod factor) and producing non-sulfated Nod factors, loose ability to nodulate alfalfa and acquire ability to nodulate pea (Lerouge et al., 1990). Thus, by introduction of the host-specific genes modifying Nod factor structure into rhizobial strains it is possible to change host range of these strains and render them efficiently nodulating other host plants (Lopez-Lara et al., 1996; Ovtsyna et al., 1998; Pacios-Bras et al., 2000).

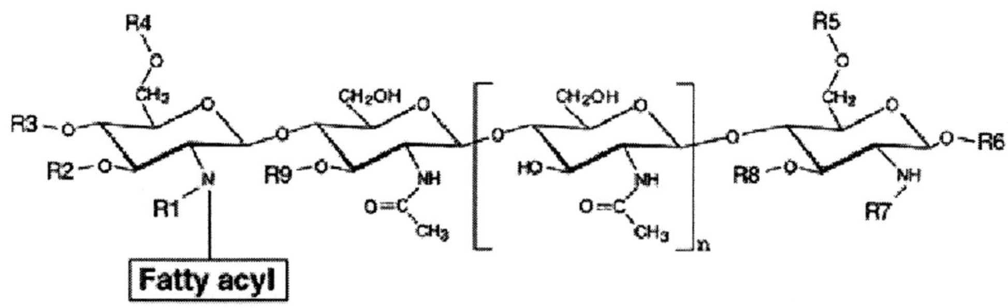

Figure 1. General structure of rhizobial Nod factors (Spaink, 2000).

There is a number of evidence that Nod factor might act in legumes by changing the internal plant hormone balance (Hirsh et al., 1994). Classical plant hormones like auxines, cytokinins, ethylene, abscisic acid, gibberellic acid has shown to influence nodule development (Hirsh et al., 1989; de Billy et al., 2001; Penmetsa and Cook, 1997; Heidstra et al., 1997). In addition, signaling compounds including peptide hormones, salicylic acid, jasmonic acid, uridine, falvonoids, reactive oxyden species, niric oxide, nitrates, also has an effect on nodulaiton (Mathesius et al., 1998; Ferguson and Mathesius, 2003).

Table 1. Host specificity of *Rhizobium* – legume interaction and Nod factor structure.

Host plant	Nodule bacteria	Substitutions on Nod factor backbone
Alfalfa, *Melilotus*, *Trigonella*	*Sinorhizobium meliloti*	$C_{16:2}$, R4=Ac, R5=S
Pea, vetch, lens, *Lathyrus*	*Rhizobium leguminosarum* bv. *viciae*	$C_{18:1}$, $C_{18:4}$, R4=Ac
Clover	*Rhizobium leguminosarum* bv. *trifolii*	$C_{18:0}$, $C_{18:1}$, $C_{20:3}$, $C_{20:4}$, R4=Ac
Soybean, peanut, siratro	*Bradyrhizobium japonicum*	$C_{16:0}$, $C_{18:0}$, $C_{18:1}$ R5=MeFuc
Lotus, lupin, siratro	*Mezorhizobium loti*	$C_{18:0}$, $C_{18:1}$, R2=Cb, R5-Fuc, AcFuc
Acacia	*Rhizobium teranga* bv. *acaciea*	$C_{16:0}$, $C_{18:0}$, $C_{18:1}$, R1=Me, R2=Cb, R4=Cb, R5=S, H
Sesbania	*Azorhizobium caulinodans*	$C_{18:1}$, $C_{16:0}$, R1=Me, R2=OH, R4=Cb, H, R5=Fuc, R8=Ara
112 legume genera (*Desmodiinae, Phaseoleae*), non-legume *Parasponia*	*Rhizobium sp.* strain 234	$C_{16:1}$, $C_{18:0}$, $C_{18:1}$, R1=Me, R2=Cb, OH, R3=Cb, H, R5=MeFuc, AcMeFuc, SmeFuc, R9=OH

Abbreviations: R1 – R9 – chemical substitutions, if not indicated, R=hydrogen. $C_{18:0}$, $C_{16:2}$ etc. – fatty acid structure; Ac, acetyl; Ara, arabinosyl; Cb, carbamoyl; Fuc, fucosyl; H, hydrogen; Me, methyl; S, sulfate; MeFuc, methylfucose; AcMeFuc, acetylmethylfulcose; SMeFuc, sulfated methylfucose

The structure of rhizobial Nod factors is very close to the structure of plant defense response elicitors - chitin oligomers. These molecules are released from chitin-containing organisms, such as fungi and invertebrates, during pathogen attack and induce plant genes implicated in the synthesis of phytoalexins, reactive oxygen molecules and hydrolytic enzymes. Early reactions of plants to chitin oligomers and Nod factors seem to be very similar: they both induce plasma membrane depolarization and changes in Ca^{2+}, K^{+} and anion concentrations (Felle et al., 2000). In suspension cultures of various plant species Nod factors and chitin oligomers have been shown to induce similar responses including alkalinisation, reactive oxygen production and induction of an isoflavonreductase gene (Staehelin et al., 1994; Müller at al., 2000; Day et al. 2001). These data suggest that, at least in some cases, perception systems for chitin oligomers and Nod factors might interfere with each other. However, the further reactions leading to contrast (defence or symbiotic) responses must be considerably different. Thus, the way of how legumes discriminate Nod factors and chitin oligomers signals to induce symbiotic and defence responses and how transduction of these signals proceeds, still has to be studied.

Endogenous Nod Factor - Like Molecules

Since Nod factors possess morphological activity in plants and act in nanomolar concentrations, it is an intriguing question whether such molecules are present in plant tissues and might play a role of endogenous plant growth regulators (Spaink, 1996; van der Holst et al., 2001). Indeed, introduction of the *nodA* and *nodB* genes involved in Nod factor synthesis in tobacco results in severe changes in development and morphology of plants (Schmidt et al., 1993). Adding of Nod factors or 32-kD endochitinase (presumably generating Nod factor analogs from other plant molecules) restore normal embryo development in temperature – sensitive somatic embryo carrot mutants, indicating that Nod factor – like molecules are crucial for plant embryogenesis (de Jong et al., 1993). Chitinase – modified N-acetylglucosamine and glucosamine – containing arabinogalactan proteins are also able to control somatic embryogenesis in carrot (van Hengel et al., 2001). Recently it was demonstrated that endogenous Nod factor – like molecules promote early somatic embryo development in Norway spruce (Dyachok et al., 2002). All these data favor hypothesis about involvement of endogenous Nod factor-like molecules in plant development. Moreover, molecules with similar structure and functions seem to be present also in animal tissues (Spaink, 1996; van der Holst et al., 2001). Thus, rhizobial protein NodC, responsible for synthesis of Nod factor backbone, has high similarity with DG42 protein, active in early embryogenesis of frog, fish and mouse (Rosa et al., 1988; Semino et al., 1996). This protein works *in vitro* as chitinoligosaccharide-synthase (Semino et al., 1995; 1996). Microinjection of the DG42 antiserum or the *Bradyrhizobium* NodZ enzyme in fertilized eggs of zebrafish led to severe defects in trunk and tail development (Bakkers et al., 1997; Semino et al., 2000). Thus, signal molecules similar to rhizobial Nod factors might be widespread in living organisms and play an important role in development of plants and vertebrates.

Possibility of Practical Application of Nod Factor – Like Signal Molecules

Nod factors are absolutely required for nodulation. Addition of purified Nod factor molecules has a positive effect both on *Rhizobium* – legume symbiosis and plant interaction with mycorrhizal fungi and is in some cases sufficient for nodule formation (Relic et al., 1994; Xie et al., 1995; Broughton et al., 2002). However, large – scale purification of these molecules from bacterial strains required for their commercial production is too expensive. In addition, Nod factors are quickly degraded to non-active products in rhizosphere by plant hydrolytic enzymes (Heidstra et al., 1994; Staehelin et al., 1995; Ovtsyna et al., 2000). Nevertheless, these molecules are good candidates for future practical use as plant growth regulators: it was recently shown that application of the Nod factor of *Bradyrhizobium japonicum* enhanced germination of a variety of economically important non-legume plants belonging to diverse botanical families: *Zea mays*, *Oryza sativa* (*Poaceae*), *Beta vulgaris* (*Chenopodiaceae*), *Glycine max*, *Phaseolus vulgaris* (*Fabaceae*), and *Gossypium hirsutum* (*Malvaceae*), under laboratory, greenhouse and field conditions. LCO also enhanced the early growth of corn seedlings under greenhouse conditions (Prithiviraj et al., 2003). Stimulating effect of Nod factors on non-legumes might be explained by the fact that root extracts from non-legume plants are able to activate CO production by *B.japonicum* (Lian et al., 2002). These findings suggest the possible use of LCOs for improved crop production.

The problem of large - scale production of Nod factor – like molecules might be solved by their chemical synthesis. Synthesis of lipochitooligosaccharides (LCOs) was performed by several groups. Namely, LCO analogs of *Sinorhizobim meliloti*, *Bradyrhizobium japonicum*, *Rhizobium leguminosarum* bv. *vicieae* Nod factors were produced (Table 2). LCO derivatives labeled with different fluorescent groups were also reported. Fluorescently labeled molecules are currently used for their visualization during interaction with plant, in search for the plant receptors to Nod factors and in study of related signal transduction mechanisms (Gadella et al. 1997; Goedhart et al., 1999, 2000; Cullimore et al., 2001; Goedhard et al., 2003).

Table 2. Chemically synthesized Nod factor analogs.

Nod factor	Rhizobia	Reference
NodRm –IV ($C_{16:2}$) NodRm –IV (Ac, S, $C_{16:2}$) NodRm –IV (S, $C_{16:2}$) NodRm –IV (Ac, $C_{16:2}$)	*Sinorhizobium meliloti*	Nicolaou et al., 1992
NodRm –IV ($C_{16:2}$, S)	*Sinorhizobium meliloti*	Ikeshita et al., 1994a
NodBj-V (RCO, MeFuc) NodBj-IV(RXO, MeFuc)	*Bradyrhizobium japonicum*	Ikeshita et al., 1994b
NodBj-IV ($C_{18;1}$, Fuc, Gro)	*Bradyrhizobium japonicum*	Ikeshita et al., 1995
NodRm –IV (S, $C_{8:1}$) NodRm –IV (S, $C_{12:1}$) NodRm –IV (S, $C_{16:1}$) NodRm –IV (Ac, $C_{18:1}$) NodRm –IV (Ac, $C_{18:2}$)	*Sinorhizobium meliloti*	Demont-Caulet et al., 1999
NodRlv –IV (BODIPY FL $_{480/515}$-C_{16})	*Rhizobium leguminosarum bv. viciae*	Gadella et al., 1997
NodRlv –IV (BODIPY$_{558/568}$-C_{12}) NodRlv –IV (BODIPY$_{581/591}$-C_{11}) NodRlv –IV (BODIPY$_{581/591}$-C_{16})	*Rhizobium leguminosarum bv. viciae*	Goedhart et al., 1999
NodSm –IV (S, BODIPY$_{558/568}$-C_{12}) NodSm –IV (S, BODIPY$_{581/591}$-C_{11}) NodSm –IV (S, BODIPY$_{581/591}$-C_{16})	*Sinorhizobium meliloti*	Goedhart et al., 2003

Here we describe newly synthesized Nod factor analogs, carrying new type of fluorophore. We have analyzed biological activity of these compounds by testing their ability to induce early Nod factor – related reactions.

MATERIAL AND METHODS

LCOs

Five chemical analogous of rhizobial lipochitooligosaccharide signal molecules were synthesized in the Institute of Organic Chemistry, Moscow, Russia, in group of Dr. Torgov. They are listed below and their structure is shown on Figure 2.

LCO1: II ($C_{16:0}$, Ac Me Fuc)
LCO2: II ($C_{16:0}$, Me Fuc)
LCO3: II ($C_{11:0}$, fluo $_{520/590}$, Ac Me Fuc)
LCO4: II ($C_{11:0}$, fluo $_{520/590}$, Me Fuc)
LCO5: IV ($C_{16:0}$, Ac Me Fuc on 3^{d} GlcNAc)

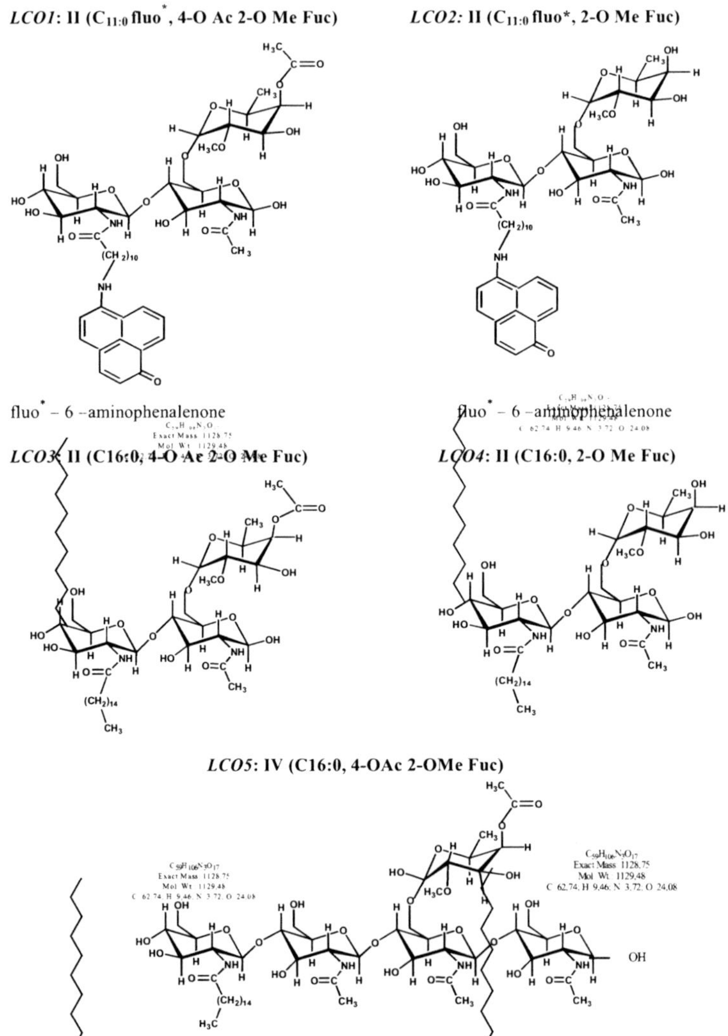

Figure 2. Structure of chemically synthesized Nod factors (this work).

All these compounds carry 2-O methylfucose (substituted with 4-O acetyl group on LCO1, 3, 5) at the reducing terminal N-acetylglucosamine and fatty acid chain at the non-reducing terminus – vaccenic acid ($C_{16:0}$, LCO1, 2, 5) or reduced $C_{11:0}$ chain, if the fluorescent label is attached to the fatty acid (LCO 3, 4). Four compounds are fucosylated dimers of N-acetylglucosamine and one is tetramer having Ac Me Fuc group at the third GlcNAc (LCO5) (Figure 3). Synthesized fluorescent group (6 –aminophenalenone) has an excitation at 520 nm and emission at 590 nm. Stock solutions of LCOs (1 mM) were prepared either in 30% ethanol or in DMSO. For biological assays they were diluted to 10^{-6} – 10^{-8} M in water.

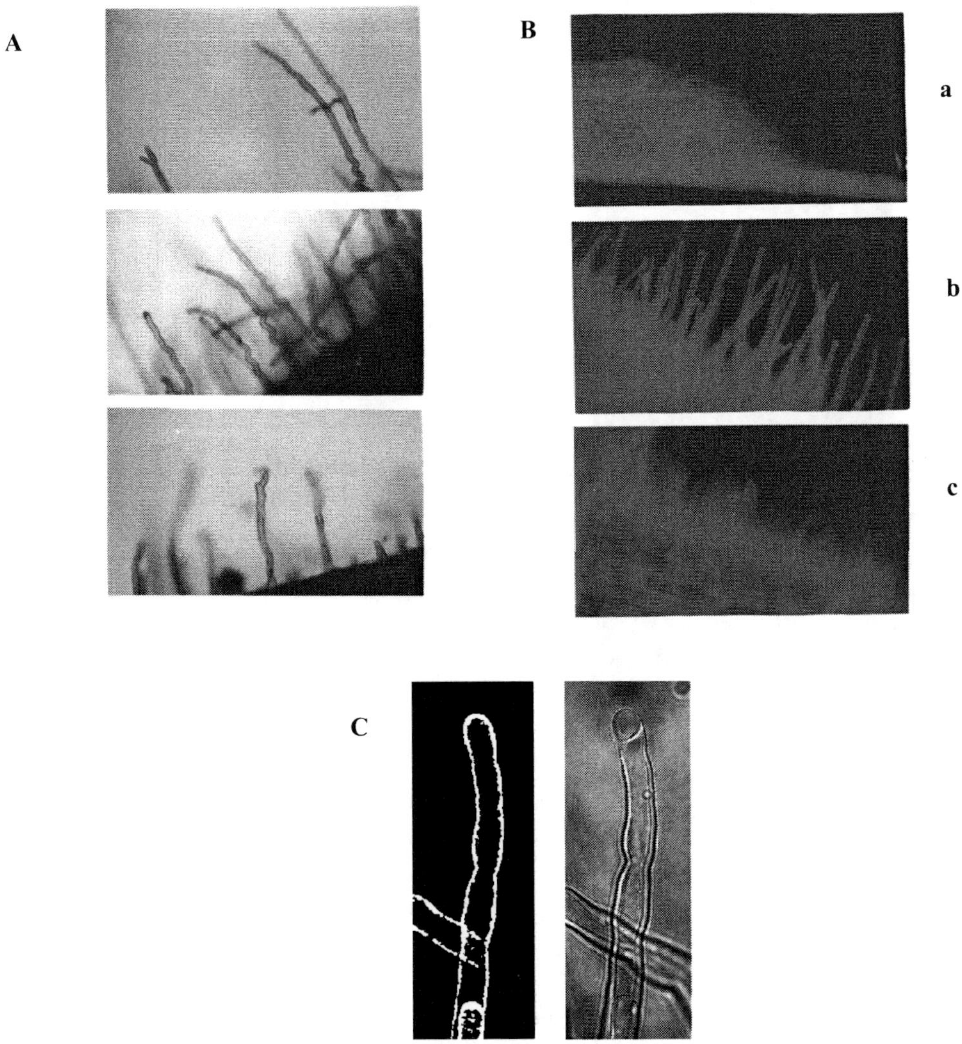

Figure 3. Effects of synthetic LCOs on plant roots. **A** – root hair deformations induced by application of 10^{-6} M LCO1 on roots of **a**), **b**) *Lotus*, **c**) *Glycine*. **B** - microscope image of *Lotus* roots incubated with **a**) sand particles coated by LCO1 (10^{-4} M), **b**), **c**) with solution of LCO1 (10^{-6} M). **C**- localization of LCO1 (10^{-7} M) on Medicago truncatula roots.

Plant Material and Bacterial Strains

Seeds of *Lotus corniculatus*, *Glycine gracilis* and *Vigna unguiculata* were kindly provided by All Russia Institute for Plant Breeding (St.-Petersburg, Russia). They were surface-sterilized and germinated for 3 days on 1% agar plates. For root hair deformation assay (HAD) they were transferred either on Petri dishes (*Lotus*) or into glass tubes (*Glycine*, *Vigna*) containing Jenssen medium with 1% agar (van Brussel et al., 1982). LCOs or bacteria were applied to root close to root tip either as a drop of solution (10^{-6} – 10^{-7}M) or on sand particles coated with LCOs (10^{-4}M) (Lopez-Lara et al., 1996). Root hair deformations were scored under microscope (Opton, Zeiss, Germany) 24-48 h after LCO/bacteria application.

Rhizobial strains efficiently nodulating *Lotus*, *Glycine* and *Vigna* were obtained from the collection of All Russia Research Institute for Agricultural Microbiology. Their ability to form nodules on their host plants was tested *in vitro* and optimal conditions for nodulation test and for HAD assay were developed.

Results and Discussion

Analysis of the Biological Activity of Synthetic LCOs on Roots of Model Legumes

Synthetic LCOs have been tested for biological activity in root hair deformation assay (Table 3). To estimate biological activity of these synthetic LCOs, a set of definite legume plants was chosen. Methylated and acetylated fucosyl residues are present on Nod factors of rhizobia nodulating cowpea (*Vigna unguiculata*), soybean (*Glycine max, Glycine soja*) and *Lotus ssp.* (Schmidt et. al., 1994; Gehring et.al., 1997; Lopez-Lara et.al., 1996). Therefore, the appropriate plant species were used for our experiments. All the tested molecules demonstrated week biological activity in comparison to symbiotic bacteria. The most active compounds were dimeric LCOs carrying O-acetyl group at the fucosyl residue (LCO1, 3) - they elicited up to 10 – 15 % of root hair deformation on *Vigna* and *Glycine* and 2-4 % on *Lotus* (2-4 folds more than in control plants) (Table 3). Thus, acetylation of fucose was very important for activity of LCOs on the tested legume plants. No activity was found for tetrameric LCO, carrying acetylfucose group at the third GlNAc. Since majority of indigenous Nod factors carrying fucosyl modification have it at the reducing terminal N-acetylglucosamine, it seems that location of modified fucosyl residue exactly at the reducing terminus is crucial for its biological activity. To verify whether synthetic LCOs, carrying 2-O-acetyl, 4-O-methylfucose at the reducing end, demonstrated real activity, we used the legume plant *Medicago truncatula* required rhizobia producing LCOs with other modifications (sulfate group at the reducing end). No activity was found for dimeric LCO on *Medicago truncatula* (data non shown), what is in the accordance with our suggestion.

Analysis of the fluorescently labelled LCOs under microscope has shown that LCOs adsorb to the plant roots (*Lotus spp.)* in a short period and are fixed presumably to the plant cell wall. Fluorescence of the new label (6-aminophenalenone, exitation at 520 nm and emission at 590 nm) is clearly visible as a bright red spot at site of LCO application on the almost black background of root autofluorescence (Fig. 4). Root hairs also acquire red fluorescence and are able to grow and deform after incubation with labelled LCOs (Fig. 4).

Thus, fluorescent label attached to LCOs was well distinguished on the root and did not reduce their biological activity with responsive plants. At the same time fluorescently labelled molecules were well visible in the cell wall of *Medicago truncatula* root hairs, although no biological activity for tested LCO was demonstrated on this plant. LSOs binding to the root hair cell wall is not strictly correlated with bioactivity. Thus, immobilisation in the cell wall might be important for increasing the efficiency of initial LCOs perception by plant cells.

Table 3. Activity of the synthetic LCOs in root hair deformation (Had) assay.

Plant species	LCOs	Had, % *	Notes
Vigna unguiculata	Control	1.8 ± 0.74	twisted long hairs
	II Me Ac Fuc fluo (1)	**11.2 ± 3.68**	
	II Me Fuc fluo (2)	-	
	II Me Ac Fuc (3)	**7.7 ± 5.36**	
	II Me Fuc (4)	-	
	IV Me Ac Fuc (5)	2.9 ± 1.33	
	bacterial strain 0502	**30.0 ± 14.0**	
Lotus corniculatus	Control	0.45 ± 0.26	Root hair branching are quite often in treated variants
	II Me Ac Fuc fluo (1)	**2.5 ± 0.58**	
	II Me Fuc fluo (2)	0.7 ± 0.24	
	II Me Ac Fuc (3)	**3.4 ± 1.87**	
	II Me Fuc (4)	1.8 ± 0	
	IV Me Ac Fuc (5)	0.8 ± 0.33	
	bacterial strain 1801	**35.9 ± 3.87**	
Glycine gracilis	Control	7.8 ± 0.95	A lot of deformations in control (twisted, spiral, thick apexes)
	II Me Ac Fuc fluo (1)	**13.7 ± 2.13**	
	II Me Fuc fluo (2)	4.0 ± 0.75	
	II Me Ac Fuc (3)	5.2 ± 0.38	
	II Me Fuc (4)	-	
	IV Me Ac Fuc (5)	-	
	bacterial strain 24102	**39.2 ± 1.75**	

*Results are presented as average ± standard error from 2 experiments, 4-5 plant per each treatment.

First designed fluorescent Nod factors had a green fluorescent label NodRlv –IV (BODIPY FL $_{480/515}$-C_{16}). They possessed high biological activity but their spectral characteristics were not good enough for microscopy study, since at these excitation and emission wavelengths high root autofluorescence was also observed (Gadella et al., 1997). Next generation of labeled Nod factors, carried label BODIPY$_{558/568}$-C_{12} with a shift of emission to red light, was better suitable for microscope observation (Goedhart et al., 1999, 2000). Fluorescent group attached to Nod factor analogs used in this study also has emission in red region, making them excellent for microscopic investigation. This new label is less spacious than BODIPY label and seems not to affect biological activity of Nod factors. We suppose that Nod factor analogs tested in this work, and their longer derivatives which will be produced in future, will be well suitable for study of Nod factor binding and perception by plants.

The high specificity and the low concentration at which Nod factors are active strongly suggest the involvement of a Nod factor receptor (Cullimore et al., 2001). Analysis of legume mutants that can not respond to rhizobial Nod factors and do not form nodules, has led to the identification of genes encoding potential receptors for Nod factors - *sym10* in pea (Walker et. al., 2000), NFP in *Medicago truncatula* (Ben Amor et. al., 2003) and NFR1, NFR5 in *Lotus japonicus* (Radutoiu et.al., 2003). Recently related genes have been cloned and it was shown to encode receptor like kinase proteins (RLK) containing LysM domains in the extracellular part. Similar LysM domains were identified in bacterial proteins (lysins) involved in peptidoglycans binding. In eukaryotes, LysM domains are found in chitinooligosaccharide binding domains of some chitinases. These results demonstrate an important role for Lys-RLKs in initial Nod factor perception and create a good basement for deciphering of mechanisms of symbiosis specificity.

At the same time, experiments with purified Nod factors and bacterial mutants have shown that specificity of plant responses highly depends on LCOs chemical structure. It suggests that recognition of LCOs on the host plant is accomplished by several receptors that differ from one another in stringency of recognition and may trigger different response pathways (Journet et al., 1994; Felle et. al., 1998; Minami et. al., 1996). In legumes the structural requirements for Nod factors may be more stringent at the stage of bacteria entry in root-hair cell than for induction of earlier symbiotic responses. It suggests the existence of two types of Nod factors receptors – one for "entry" and another for "signaling" (Cullimore and Denarie, 2003). In support of this suggestion, two genes *LYK3* and *LYK4* were identified in *Medicago truncatula*, which are required for Nod-factor-dependent rhizobia infection (Limpens et.al., 2003). These genes encode proteins, which also belong to class of LysM-RLKs. The similar gene *sym2* in pea permits infection only by rhizobia producing Nod factors with specific modifications (Geurts et. al., 1997). Thus, the characterized *LYK3/LYK4* and *sym2* genes are highly probable candidates for role of "entry" receptors in legume plants.

The existence of several potential perception sites for Nod factors was shown using biochemical approach. Two different Nod factor binding sites were characterized in root extracts and cell cultures of *Medicago truncatula* (Bono et al., 1995; Cullimore, 2001). A low affinity binding site, NFBS1, was found in root extracts of *M.truncatula* and also in tomato. High affinity binding site, NFBS2, was found in cell culture extracts of *M.varia*. Both binding activities do not discriminate between sulfated and non-sulfated Nod factors, suggesting the existence of another, more specific Nod factors receptor in legume roots (Cullimore, 2001; Goedhard et al., 2003). Using fluorescently labeled Nod factor analogs it was demonstrated that they are accumulated and immobilized in the cell wall (Goedhard et al., 1997, 2000). This accumulation is not host – specific: both sulfated and non-sulfated Nod factors are identically immobilized in root hair cell wall of *Vicia sativa* and *Medicago truncatula* (Goedhard et al., 2003). Thus, it is hypothesized that at the first step Nod factors are anchored by non-specific receptor (probably corresponding to NFBS1 binding site, Bono et al., 1995) and than they are transduced to the cytosol by binding to more specific Nod factor signaling receptor (Goedhard et al., 2003). Our results are in good agreement with these observations. Although the tested fluorescent Nod factors analogs are significantly shorter than wild-type molecules and substitutions on their backbone are also different from those of native compounds, they are able to absorb on plant cell wall, probably due to interaction with common non-specific receptor.

Chemically synthesized Nod factor analogs, available in various structural modifications and in unlimited amounts, should make a significant contribution to our understanding of the perception and signal transduction of chitin oligosaccharides and Nod factors. Oligosaccharides are now acknowledged as probably the most important class of signalling and recognition molecules (Rudiger et al., 2000) due to their superior information encoding capacity (compared to proteins and nucleic acids). The availability of structurally related compounds triggering contrasting plant responses (symbiosis or defence) thus provides an excellent model to study their recognition. These molecules could have biotechnological applications in promoting environmentally-friendly, sustainable, agricultural systems with low chemical inputs. Biopreparations based on chitin and chitosane show good results in inhibition of pathogens and increasing of total crop productivity (Hadwiger et al., 1994; Flese and Panda, 1999; Tsigos et al., 2000). For legumes, their key role in sustainable agricultural systems is well-known and there is an increasing need to improve the efficiency of nodulation of these plants in order to promote biological nitrogen fixation and hence to reduce the input of costly and polluting nitrogenous fertilisers (Sanchez et al., 1999). The studies on Nod factor perception using synthetic Nod factor analogs may be exploited in order to provide improved rhizobial strains and legume genetic material to intensify nodulation efficiency.

ACKNOWLEDGMENTS

This work was supported by NWO, Russian Ministry of Science and Technologies and CRDF ST-012-0 grants.

REFERENCES

[1] Bakkers J., Semino C.E., Stroband H. et al.: An important developmental role for oligosaccharides during early embryogenesis of cyprinid fish. *Proc. Nat. Acad. Sci. USA*, **94**, 7982 (1997).

[2] Ben Amor B., Shaw S., Oldroyd J., Maillet F., Varma Penmetsa R., Cook D., Long S., De′ Narie J., Gough C.: The NFP locus of *Medicago truncatula* controls an early step of Nod factor signal transduction upstream of a rapid calcium flux and root hair deformation. *The Plant Journal,* **34**, 495 (2003).

[3] Bono J.-J., Riond J., Nicolaou K.C., Bockovich N.J., Estevez V.A., Cullimore J.V., Ranjeva R.: Characterisation of a binding site for chemically-synthesized lipo-oligosaccharidic NodRm factors in particulate fractions prepared from roots. *Plant J.*, **7,** 252 (1995).

[4] Broughton W.J., Zhang F., Staehelin C.: Signals exchanged between legumes and Rhizobium: agricultural uses and perspectives. *Plant and Soil,* 1, (2002).

[5] Cardenas L., Holdaway-Clarke T.L., Sanchez F., Quinto C., Feijo J.A., Kunkel J.G., Hepler P.K.: Ion changes in legume root hairs responding to Nod factors. *Plant Physiology,* **123**, 443 (2000).

[6] Cullimore, J., Ranjeva, R. And Bono. J-J.: Perception of lipo-chitooligosaccharidic Nod factors in legumes. *Trends in Plant Sci.,* **6**, 24 (2001).

[7] Cullimore J. and Denarie J.: How legumes select their sweet talking symbionts. *Science*, **302**, 575 (2003).

[8] Day R.B., Okada M., Ito Y., et al.: Binding site for chitin oligosaccharides in the soybean plasma membrane. *Plant Physiology*, **126**,1162 (2001).

[9] De Billy F., Grosjean C., May S., Bennett M., Cullimore J.V.: Expression studies on AUX1-like genes in Medicago truncatula suggest that auxin is required at two steps in early nodule development. *Mol Plant Microbe Interacts*, **14**, 267 (2001).

[10] De Jong A.J., Heidstra R., Spaink H.P. et al.: Rhizobium Lipooligosaccharides Rescue a Carrot Somatic Embryo Mutant. *Plant Cell.*, **5**, 615 (1993).

[11] Demont-Caulet N., Mailett F., Tailler D., Jacquinet J.-C., Prome J.-C., Nicolaou K.C., Truchet G., Beau J,-M., Denarie J.: Nodule-inducing activity of synthetic Sinorhizobium meliloti nodulation factors and related lipo-chitooligosaccharides on alfalfa. Importance of the acyl chain structure. *Plant Physiology*, **120,** 83 (1999).

[12] Dénarié, J., Debellé, F., Promé, J.-C.: Rhizobium lipo-chitooligosaccharide nodulation factors: signaling molecules mediating recognition and morphogenesis. *Annu. Rev. Biochem.,* **65**, 503 (1996).

[13] Dyachok J.V., Wiweger M., Kenne L., Von Arnold S.: Endogenous Nod-Factor-Like Signal Molecules Promote Early Somatic Embryo Development in Norway Spruce. *Plant Physiology,* **128**, 523 (2002).

[14] Endre G., Kereszt A., Kevei Z., Mihacea S., Kalo P., Kiss G.: A receptor kinase gene regulating symbiotic nodule development. *Nature* **417** (27), 962 (2002).

[15] Felle H.H., Kondorosi E., Kondorosi A., Schultze M.: The role of ion fluxes in Nod factor signaling in *Medicago sativa* L. *Plant Journal,* **13,** 455 (1998).

[16] Felle H.H., Kondorosi E., Kondorosi A., Schultze M.: How alfalfa root hairs discriminate between Nod factors and oligochitin elicitors. *Plant Physiol.,* **124** 1373 (2000).

[17] Journet E.P., Pichon M., Dedieu A., De Billy F., Truchet G., Barker D.G.: Rhizobium meliloti Nod factors elicit cell-specific transcription of the ENOD12 gene in transgenic alfalfa. *Plant J.*, **6**, 241 (1994).

[18] Ferguson B.J. And Mathesius U.: Signaling interactions during nodule development. *J Plant Growth Regul.,* **22**, 47 (2003).

[19] Flese A.P., Panda T.: Regulation and cloning of microbial chitinase genes. *Appl Microbiol Biotechnol.*, **51**(2), 141 (1999).

[20] Gadella T.W.J., Jr., Vereb G., Hadri A.-E. et al.: Microspectroscopic imaging of nodulation factor – binding sites on living Vicia sativa roots using a novel bioactive fluorescent nodulation factor. *Biophys.* J., **72**, 1986 (1997).

[21] Geurts R., Heidstra R., Hadri A-E., Downie J.A., Franssen H., Van Kammen A., Bisseling T.: *Sym2* of pea is involved in a nodulation factor-perception mechanism that controls the infection process in the epidermis. *Plant Physiology*, **115**, 351 (1997).

[22] Goedhart J., Rohrig H., Hink M.A., Van Hoek A., Visser A.J.W.G., Bisseling T., Gadella T.W.J.: Nod factors integrate spontaneously in biomembranes and transfer rapidly between membranes and to root hairs but transbilayer flip-flop does not occur. *Biochemistry*, **38,** 10898 (1999).

[23] Goedhart J., Hink M.A., Visser A.J.W.G., Bisseling T., Gadella T.W.J.: In vivo fluorescence correlation microscopy (FSM) reveals accumulation and immobilization of Nod factors in root hair cell wall. *The Plant J.*, **21**(1), 109 (2000).

[24] Goedhart J., Bono J.-J., Bisseling T., Gadella T.W.J.: Identical accumulation and immobilization of sulfated and non-sulfated Nod factors in host and nonhost root hair cell walls. *Mol. Plant - Microbe Interact.,* **16** (10), 884 (2003).

[25] Hadwiger L.A., Ogawa T., Kuyama H.: Chitosan polymer sizes effective in inducing phytoalexin accumulation and fungal suppression are verified with synthesized oligomers. *Molecular Plant-Microbe Interact.*, **7**, 531 (1994).

[26] Heidstra, R., Geurts, R., Franssen, H., Spaink, H. P., Van Kammen, A., and Bisseling, T.: Root hair deformation activity of nodulation factors and their fate on Vicia sativa. *Plant Physiol.*, **105**, 787 (1994).

[27] Heidstra R., Yang W.C., Yaicin Y., Peck S., Emons A.M., Van Kammen, A., and Bisseling, T.: Ethylene provides positional information on cortical cell division but is not involved in Nod factor – induced root hair tip growth in Rhizobium – legume interaction. *Development* **124,** 1781 (1997).

[28] Hirsch A.M., Bhuvaneswari T.V., Torrey J.G., Bisseling T.: Early noduline genes are induced in alfalfa root outgrowths elicited by auxin transport inhibitors. *Pros. Natl. Acad. Sci.*, **86**, 1244 (1989).

[29] Hirsh A.M., Fang Y.: Plant hormones and nodulation: what's the connection? *Plant Mol Biol.,* **26**(1), 5 (1994).

[30] Hirsh A.M., Lum M.R., Downie, J.A.: What makes the rhizobia-legume symbiosis so special? *Plant Physiol.,* **127**(4),1484 (2001).

[31] Ikeshita S, Nakahara Y, Ogawa T.: Synthesis of lipooligosaccharide nodulation signals NodBj-V(RCO, MeFuc) and NodBj-IV(RCO, MeFuc) of Bradyrhizobium japonicum. *Glycoconj J.,* **11**(3), 257 (1994).

[32] Ikeshita S, Nakahara Y, Ogawa T.: Synthetic studies on the lipooligosaccharide Nod Bj-IV (C18:1, Fuc, Gro) produced by Bradyrhizobium japonicum strain USDA61.*Carbohydr Res.,* **266**(2),1 (1995).

[33] Lian B., Souleimanov A., Zhou X., Smith D.L.: In vitro induction of lipo-chitooligosaccharide production in Bradyrhizobium japonicum cultures by root extracts from non-leguminous plants. *Microbiol Res.,* **157**(3), 157(2002).

[34] Limpens E., Franken C., Smit P., Willemse J., Bisseling T., Geurts R.: LysM domain receptor kinases regulating rhizobial Nod factor-induced infection. *Science*, **302,** 630 (2003).

[35] Lerouge P., Roche P., Faucher C., Maillet F., Truchet G., Prome J.-C., Denarie J.: Symbiotic host- specificity of Rhizobium meliloti is determined by a sulfated and acylated glucosamine oligosaccharide. *Nature*, **344**, 781 (1990).

[36] Lopez-Lara I.M., Blok-Tip L., Quinto C., Garcia M.L., Stacey G., Bloemberg G., Lamers G.E.M., Lugtenberg B.J.J., Thomas-Oates J.E., Spaink H.P.: NodZ of Bradyrhizobium extends the nodulation host range of Rhizobium by adding a fucosyl residue to nodulation signals. *Mol. Microbiol.*, **21**(2), 397 (1996).

[37] Mathesius, U., Schlaman H.R.M., Spaink H.P. et al.: Auxin transport inhibition precedes nodule formation in white clover roots and is regulated by flavonoids and derivatives of chitin oligosaccharides. *Plant J.* **14**(1), 23 (1998).

[38] Minami E., Kouchi H., Cohn J.R., Ogawa T., Stacey G.: Expression of the early nodulin, *ENOD40*, in soybean roots in response to various lipo-chitin signal molecules. *Plant Journal,* **10**, 23 (1996).

[39] Muller J., Staehelin C., Xie Z.P., Neuhaus-Url G., Boller T.: Nod factors and chitooligomers elicit an increase in cytosolic calcium in aequorin-expressing soybean cells. *Plant Physiol.*, **124**, 733 (2000).

[40] Nicolaou K.C., Bockovich N.J., Carcanague D.R., Hummel C.V., Even L.F. Total synthesis of NodRm-IV factors, the *Rhizobium* nodulation signals. *J. Am. Chem. Soc.*, **114**, 8701 (1992).

[41] Oldroyd, G.E.D., Engstrom E.M., Long S.R.: Ethylene inhibits the Nod factor signal transduction pathway of Medicago truncatula. *Plant Cell,* **13,** 1835 (2001).

[42] Ovtsyna A.O., Geurts R., Bisseling T., Lugtenberg B.J.J., Tikhonovich I.A., Spaink H.P.: Restriction of host range by the sym2 allele of Afghan peas is non - specific for the type of modification at the reducing terminus of nodulation signals. *Mol. Plant - Microbe Interact.*, **11**(5), 418 (1998).

[43] Ovtsyna A.O., Schultze M., Tikhonovich I.A., Spaink H.P., Kondorosi E., Kondorosi A., Staehelin C.: Nod factors of Rhizobium leguminosarum bv. viciae and their fucosylated derivatives stimulate a Nod factor cleaving activity in pea roots and are hydrolyzed in vitro by plant chitinases at different rates. *Mol. Plant - Microbe Interact.*, **13**(8), 799 (2000).

[44] Pacios-Bras C., Jorda M.A., Wijfjes A.H., Harteveld M., Stuurman N., Thomas-Oates J.E., Spaink H.P.: A Lotus japonicus nodulation system based on heterologous expression of the fucosyl transferase NodZ and the acetyl transferase NolL in Rhizobium leguminosarum. *Mol Plant Microbe Interact.*, **13**(4), 475 (2000).

[45] Pemnetsa R.V., Cook D.R. A legume ethylene-insensitive mutant hyperinfected by its Rhizobium symbiont. *Science*, **275,** 527 (1997).

[46] Perret, X., Staehelin, C., Broughton, W.J.: Molecular basis of symbiotic promiscuity. *Microbiol Mol Biol Rev.*, **64**, 180 (2000).

[47] Prithiviraj B, Zhou X, Souleimanov A, Kahn Wm, Smith Dl.: 2003. A host-specific bacteria-to-plant signal molecule (Nod factor) enhances germination and early growth of diverse crop plants. *Planta*, **216**(3), 437 (2000).

[48] Radutoiu S., Madsen L.H., Madsen E., Felle H., Umehara Y., Grenlund M., Sato S., Nakamura Y., Tabata S., Sandal N., Stougaard J.: Plant recognition of symbiotic bacteria requires two LysM receptor-like kinases. *Nature*, **425**, 585 (2003).

[49] Relic B., Falmont F., Kopsinska J., Golinowsky W., Prome J.-C., Broughton W.J.: Biological activity of Rhizobium sp NGR234 Nod factors on Macroptilium atropurpureum. *Mol Plant Microbe Interact.*, **5**(6), 764 (1994).

[50] Roche P., Debelle F., Mailett F. et al.: Molecular basis of symbiotic host specificity in Rhizobium meliloti: nodH and nodPQ genes encode the sulfation of lipo-oligosaccharide signals. *Cell*, **67**, 1131 (1991).

[51] Rosa F., Sargent T.D., Rebbert M.L. et al.: Accumulation and decay of DG42 gene products follow a gradient pattern during Xenopus embryogenesis. *Develop. Biol.*, **129,** 114 (1988).

[52] Rudiger H., Siebert H.C., Solis D. et al.: Medicinal chemistry based on the sugar code: fundamentals of lectinology and experimental strategies with lectins as targets. *Curr. Med. Chem.*, **7** (2000).

[53] Sanchez F., Cardenas L., Quinto C.: Biological nitrogen fixation and future challenges of agriculture. The endophytic connection. *Adv. Exp. Med. Biol.*, **464,** 107 (1999).

[54] Schultze, M. And Kondorosi, A.: Regulation of symbiotic root nodule development. *Annu. Rev. Genet.* **32**, 33 (1998).

[55] Semino C.E., Robbins P.W.: Synthesis of "Nod"-like chitin oligosaccharides by the Xenopus developmental protein DG42. *Proc. Nat. Acad. Sci. USA,* **92**, 3498 (1995).

[56] Semino C.E., Specht C.A., Raimondi A., Robbins P.W.: Homologs of the Xenopus developmental gene DG42 are present in zebrafish and mouse and are involved in the synthesis of Nod-like chitin oligosaccharides during early embryogenesis. *Proc. Nat. Acad. Sci. USA*, **93**, 548 (1996).

[57] Semino C.E., Allende M.L.: Chitin oligosaccharides as candidate patterning agents in zebrafish embryogenesis. *Int J Dev Biol.* **44**(2), 183 (2000).

[58] Schmidt J., Rohrig H., John M. e. a. *Plant J.,* **4** (4), 651 (1993).

[59] Spaink, H. P., Sheeley, D. M., Van Brussel, A. A. N., Glushka, J., York, W. S., Tak, T., Geiger, O., Kennedy, E. P. Reinhold, V. N., and Lugtenberg, B. J. J.: A novel highly unsaturated fatty acid moiety of lipo-oligosaccharide signals determines host specificity of Rhizobium. *Nature*, **354**,125 (1991).

[60] Spaink, H. P.: Regulation of plant morphogenesis by lipo-chitin oligosaccharides. *Crit. Rev. Plant Sci.*, **15,** 559 (1996).

[61] Spaink, H.P. Root nodulation and infection factors produced by rhizobial bacteria. *Annu Rev Microbiol.*, **54**, 257 (2000).

[62] Staehelin C., Granado J., Muller J.: Perception of Rhizobium nodulation factors by tomato cells and inactivation by root chitinases. *Proc. Natl. Acad. Sci. USA.*, **91**, 2196 (1994).

[63] Staehelin, C., Schultze, M., Kondorosi, E., and Kondorosi, A.: Lipo-chitooligosaccharide nodulation signals from Rhizobium meliloti induce their rapid degradation by the host plant alfalfa. *Plant Physiol.,* **108**, 1607 (1995).

[64] Stokkermans T.J.W., Ikeshita S., Cohn J. et al.: Structural requirements of synthetic and natural product lipo-chitin oligosaccharides for induction of nodule primordia on Glycine soja. *Plant Physiol.*, **108,** 1587 (1995).

[65] Stracke S., Kistner C., Yoshida S., Mulder L., Sato S., Kaneko T., Tabata S., Sandal N., Stougaard J., Szczyglowski K., Parniske M.: A plant receptor-like kinase required for both bacterial and fungal symbiosis. *Nature*, **47** (27), 959 (2002).

[66] Tsigos I., Martinou A., Kafetzopoulos D., Bouriotis V.: Chitin deacetylases: new, versatile tools in biotechnology. *Trends in Biotechnology* **18**, 305 (2000).

[67] Van Brussel A.A.N., Tak T., Wetselaar A., Pees E., Wijffelman C.A.: Small Leguminosae as test plants for nodulation of *Rhizobium leguminosarum* and other Rhizobia and *Agrobacterium* harbouring a *leguminosarum* Sym plasmid. *Plant Sci lett.,* **27**, 317 (1982).

[68] Van Der Holst P.P.G., Schlaman H.R.M., and Spaink H.P.: Proteins involved in the production and perception of oligosaccharides in relation to plant and animal development. *Current Opinion in Structural Biology*, **11**, 608 (2001).

[69] Van Hengel A.J., Tadesse Z., Immerzeel P., Schols H., Van Kammen A., And De Vries S.C.: N-Acetylglucosamine and Glucosamine-Containing Arabinogalactan Proteins Control Somatic Embryogenesis. *Plant Physiology*, **125**, 1880 (2001).

[70] Walker S.A., Viprey V., Downie J.A.: Dissection of nodulation signaling using pea mutants defective for calcium spiking induced by Nod factors and chitin oligomers. *Proceedings of the National Academy of Science*, **97**, 13413 (2000).

[71] Xie Z.-P., Staehelin C., Vierheilig H., Wiemken A., Jabbouri S., Broughton W. J., Vogeli-Lange R., Boller, T.: Rhizobial Nodulation Factors Stimulate Mycorrhizal Colonization of Nodulating and Nonnodulating Soybeans. *Plant Physiol.*, **108**,1519 (1995).

In: Biotechnology in Agriculture and the Food Industry ISBN 1-59454-119-1
Editor: G. E. Zaikov, pp. 141-148

Chapter 19

THE USE OF ALTERNATIVE SIGMA FACTOR HRPL TO IMPROVE EXPRESSION OF THE TYPE III SECRETION MACHINERY OF *ERWINIA CAROTOVORA* SUBSP. *ATROSEPTICA*

T. V. Ovchinnikova, A. L. Lagonenko, A. N. Evtushenkov and Y. A. Nikolaichik

Belarusian State University, Department of Molecular Biology, Belarus, Minsk, F. Skoryna ave.,4; e-mail: nikolaichik@bsu.by

ABSTRACT

Erwinia carotovora subsp. atroseptica (Eca) is specific for temperate zone plant pathogenic bacterium which causes "black leg" of stems in potato. This property is determined by a set of different virulence factors which among others include the Avr and harpin proteins utilizing the type III secretion system (TTSS) for secretion out of the bacterial cell. To facilitate the TTSS study, we have overexpressed its master regulator, the alternative sigma factor HrpL. Overexpression of HrpL resulted in significantly increased expression of the TTSS components, which, however, had little effect on secretion of one of its substrates - the HrpW harpin.

Keywords: Erwinia carotovora, type III secretion, hrpL, hrpJ, hrpW.

Erwinia carotovora is a plant pathogenic bacterium that causes major losses of vegetable crops in temperate climates. It is also important in biotechnology as a producer of several enzymes depolymerizing plant cell walls. In recent years studies of E. carotovora have mostly concentrated on these enzymes and control of their synthesis. Secretion of the depolymerizing enzymes has also received significant attention. It is known that E. carotovora has at least three secretion systems, but until very recently most attention has being paid to two of them – type I used for secretion of proteases and type II involved in secretion of cellulases, pectate lyases and polygalacturonases (see Binet et al., 1997; Russel, 1998 for reviews of these secretion systems). One of the reasons these secretion systems have received significant attention is due to the hopes of their use for secretion of foreign proteins into the medium which could significantly ease protein purification. It appeared, however, that both type I and type II secretion systems are very substrate-specific with secretion signals thought to be encoded in the 3D structure of the C-terminus of the secreted proteins (Duong et al., 1996; Lindeberg et al., 1998). This makes the prospects of using these secretion systems for secretion of biotechnologically important enzymes uncertain, although limited success was achieved in secretion of some proteins through the type I system (Palacios et al., 2001). The type III secretion system (TTSS) seems to be much more promising in this respect. Secretion signal for the TTSS was shown to be located within the N-terminal 50 amino acids of secreted proteins (Lloyd et al., 2002). This secretion signal seems to be very generic and there were several reports of heterologous substrate secretion (Anderson et al., 1999) and unrelated proteins like neomycin phosphotransferase (Cheng et al., 2001) and adenylate cyclase (Casper-Lindley et al., 2002) being transported via TTSS when fused to this signal sequence.

The main obstacle for the wide use of the type III secretion system for secretion studies is probably its relative inefficiency and strict control. This secretion system was detected in many animal and plant pathogens, as well as in some plant-associated bacteria, and for all bacteria studied the TTSS seems to be only expressed at contact with eukaryotic host. This is not surprising as the main function of this secretory apparatus seems to be the delivery of virulence proteins directly into the cells of eukaryotic organisms (hosts for the corresponding bacterial pathogens). While some success was achieved with culturing bacteria in a medium mimicking host tissues, this approach is not very efficient as the necessary cultural conditions (low pH, minimal medium with low nitrogen concentration) appear to be at the physiological extreme for most bacteria.

Another possible approach to increase the efficiency of the TTSS is to exploit the regulatory circuits controlling its expression. Genes coding for the components of the TTSS are usually located within large cluster composed of several operons. All these operons are under control of several regulatory proteins but the most proximal to the TTSS genes is usually just one - either an AraC-type transcriptional activator or an alternative sigma factor of the ECF family, HrpL. Pseudomonas syringae and Erwinia amylovora, the bacteria whose TTSS genes are the closest to E. carotovora are known to utilize the latter type of regulator (Xiao et al., 1994;Wei and Beer, 1995) and it was recently demonstrated that hrp/hrc genes (coding for the TTSS) in E. carotovora subsp. carotovora are also under control of HrpL (Chatterjee et al., 2002).

We have previously characterized the role of TTSS in the virulence of another subspecies of E. carotovora - E. carotovora subsp. atroseptica (Eca). In our work on the TTSS in Eca we were faced with the relative inefficiency of the TTSS. We hypothesized that relatively low expression of the genes coding for the components of the TTSS may be responsible for the low amounts of the proteins secreted by this system into the cultural medium. Therefore we have decided to improve the expression of the TTSS components by employing the overproduction of the HrpL protein – the known regulator of the hrp/hrc genes. Here we report the first results of this attempt.

Materials and Methods

Bacterial Strains and Growth Conditions

Eca strain JN42 which is a rif^R r$^-$ derivative of a wild type *Eca* strain 3–2 was used for the secretion studies, *Escherichia coli* strain XL-1 Blue – for molecular cloning procedures and *E. coli* BL21(DE3) – for the production of HrpJ protein from a T7 expression vector. *Eca* and *E. coli* were routinely grown at 28° and 37°C, respectively. LB broth was used as rich, while Miller's medium A– as minimal medium, with either normal (18 mM) or reduced (5 mM) nitrogen concentration. Antibiotics were used at the following concentrations (μg/ml): ampicillin 100; gentamycin 10; rifampicin 25.

DNA Manipulations

Enzymatic reactions with DNA, including PCR were performed according to standard protocols (Ausubel *et al.*, 1992). Plasmids pUC18 and pJQ200mp19 (Quandt and Hynes, 1993) were used as cloning vectors and pET24b –for protein production. pFLAG-CTS plasmid used for making the HrpW-FLAG fusion was purchased from Sigma. Restriction endonucleases, DNA ligase, DNA extraction and DNA sequencing kits were purchased from MBI Fermentas. DNA sequencing was done on an automated DNA sequencer ALFexpress II (Amersham). Transformation of *E. coli* and *Eca* was performed by electroporation.

Plasmid Construction

pTA2 was constructed by first amplifying a 2 kb fragment of the Eca 3-2 chromosome containing the *hrpL* gene with the primers 5′CAGGAAACAGCTATGAC3′ and 5′CCAGAGCTCATATCNCTGACNACACCAATACAT3′, cloned into the pUC18 plasmid using *Sac*I and *Pst*I sites, and than subcloned into pJQ200mp19 using the same restriction sites. pFLAG-HrpW was created by amplifying the *hrpW* gene from *Eca* with the primers 5′GGAGATAAGCTTGCTGATATGACNATHAC3′ and 5′GTATTTGGCGTCGACRTTCA CYTGTARRTT3′ and fusing the amplified *hrpW* in frame with the start codon and 3' sequence coding for the Flag peptide in the pFLAG-CTS plasmid (Sigma). To express the HrpJ protein, its gene was amplified with the primers to the known *hrpJ* sequence (AJ496800) and cloned into the pET24b plasmid (Novagen).

HrpJ Purification and Antibody Production

Membrane fraction of the BL21(DE3) cells overexpressing HrpJ from the pET24-hrpJ was boiled for 15 minutes and separated via standard SDS-PAGE electrophoresis. Protein band, corresponding to HrpJ, was eluted from polyacrylamide, and about 50 μg of protein mixed with the Freund's adjuvant was used for each of the three sequential immunizations of a rabbit with two week intervals. The serum was collected a week after the last immunization and used in western blotting procedures without further purification.

Western Blotting

Standard blotting procedures were used (Ausubel *et al.*, 1992). For detection of secreted HrpW-FLAG proteins from supernatant fractions of *Eca* cultures were precipitated with trichlorous acid. HrpJ and HrpW-FLAG proteins transferred to nitrocellulose membranes were detected with polyclonal anti-HrpJ antibody prepared in the course of this work and anti-FLAG M1 monoclonal antibody (Sigma). HRP-conjugated anti-rabbit and anti-mouse secondary antibodies were purchased from Amersham.

Enhanced chemiluminescence was used for detection HRP-conjugated antibodies, chemiluminescence was recorded with the VDS-CL system and quantified using the ImageMaster 1D software (Amersham).

Results and Discussion

Molecular Cloning the *HrpL* Gene from *Erwinia carotovora* subsp. *atroseptica*

We have previously cloned and sequenced a fragment of the *hrp* cluster from *Eca* 3-2 that included *hrpJ* and the very beginning of the neighboring gene, *hrpL* (Ageichik *et al.*, 2002). This has simplified the design of primers for amplification of *hrpL*. One primer was made according to the known sequence of *hrpJ* and another was designed according to reverse translation of the most conservative peptide in the beginning of HrpX, the product of the gene we expected to be present straight after *hrpL*. PCR with this pair of primers gave the product of expected size (2 kb). This PCR product was cloned in the multicopy vector pUC18 and sequenced. Sequencing confirmed the presence of the complete *hrpL* gene in the cloned fragment (fig.1), as well as expected parts of the neighboring genes *hrpJ* and *hrpX*.

The pUC18 based construct with the *hrpL* gene was transformed into the cells of *Eca* strain 3-2. The resulting transformants appeared to have a somewhat slower growth compared to *Eca* 3-2 cells carrying the empty vector. As in our previous experience pUC18-like high copy number vectors were often quite unstable in *Eca* 3-2 cells, we have decided to subclone *hrpL* into a lower copy number vector. The pJQ200mp19 plasmid (Quandt *et al.*, 1993) based on p15A replicon was used for this purpose and restriction enzymes chosen so as to place *hrpL* under control of the *lac* promoter of the vector. The resulting construct was named pTA2.

Additional Copies of the *hrpL* Gene Increase Expression of the Type III Secretion System in *Erwinia carotovora* subsp. *atroseptica* Cells

pTA2 was transformed into *Eca* 3-2 cells and expression of the TTSS in these transformants was followed using western blotting procedure. The level of expression was measured by the amount of one of the protein components of TTSS. The HrpJ protein was chosen for this purpose as we knew that a sequence very close to the consensus of the HrpL-dependent promoter is located in front of its gene (Ageichik *et al.*, 2002). *hrpJ* is the first gene in a relatively large *hrpJ* operon and we expected that the level of its protein product should give a good indication of expression of the other TTSS components, as in *E.*

amylovora and *P. syringae* all *hrp/hrc* genes seem to be under control of HrpL. Therefore we have raised polyclonal antibodies to the HrpJ protein and used these in standard western blotting procedures to determine the amount of the HrpJ in *Eca* cells grown in various culture conditions and in the presence or absence of the pTA2 plasmid producing additional copies of the HrpL protein.

```
         10        20        30        40        50        60        70        80
----:----|----:----|----:----|----:----|----:----|----:----|----:----|----:----|
AAGGCTTGCTCTTCCGTTTTTGCCAAATGGGGCACAACAACGGAGAGTCAGCAAATGGAAATGTCTACCCTGAAACACA
M  E  M  S  T  L  K  H  I

90        100       110       120       130       140       150       160
----:----|----:----|----:----|----:----|----:----|----:----|----:----|----:----|
CGAACAGACCCCTTCGCTTTACCCAGCGTTATCTGTCGATTGGGAACAGGTGTTTCGTCAACACGGAAAVAAATTGCATA
E  Q  T  P  S  L  Y  P  A  L  S  V  D  W  E  Q  V  F  R  Q  H  G  X  K  L  H  N

170       180       190       200       210       220       230       240
----:----|----:----|----:----|----:----|----:----|----:----|----:----|----:----|
ACTTTATCCGCAAGCGCGTHAGCAACCGTGACGACGTTGAGSATTTACAGCAAATGACCTATCTGGAAGTGCTCAAACAT
F  I  R  K  R  V  S  N  R  D  D  V  E  X  L  Q  Q  M  T  Y  L  E  V  L  K  H

250       260       270       280       290       300       310       320
----:----|----:----|----:----|----:----|----:----|----:----|----:----|----:----|
CAGGATAAWTTTGCGGGAGCATCGCGGCCGGAGACGTGGGTATTCGGTATTGCCCTCAACCTGGTTCGCAATTACTCACG
Q  D  X  F  A  G  A  S  R  P  E  T  W  V  F  G  I  A  L  N  L  V  R  N  Y  S  R

330       340       350       360       370       380       390       400
----:----|----:----|----:----|----:----|----:----|----:----|----:----|----:----|
CAGGCACGCCAGCGGCTCAGCAATGGGCGATGAGATGCTGGAACATATCGCGATAGATCTCGACCCCGGCGCTATCACCG
R  H  A  S  G  S  A  M  G  D  E  M  L  E  H  I  A  I  D  L  D  P  G  A  I  T  E

410       420       430       440       450       460       470       480
----:----|----:----|----:----|----:----|----:----|----:----|----:----|----:----|
AAAGCCAGCGGGCATTGAAACGCGCGATGGATGCCATCGTCTCACTGCCTGAAGATACCCGTCAGATGCTGATGCAATTG
S  Q  R  A  L  K  R  A  M  D  A  I  V  S  L  P  E  D  T  R  Q  M  L  M  Q  L

490       500       510       520       530       540       550       560
----:----|----:----|----:----|----:----|----:----|----:----|----:----|----:----|
CTGGATACCGATGCCAGCTATCAGGATCTCGCGTYGCAGTTGGAGATCCCGATCGGTACGGTGCGGTCGCGGCTGTCTCG
L  D  T  D  A  S  Y  Q  D  L  A  X  Q  L  E  I  P  I  G  T  V  R  S  R  L  S  R

570       580       590       600       610       620       630       640
----:----|----:----|----:----|----:----|----:----|----:----|----:----|----:----|
TGCCCGAAGTGTGATTCGYCAGGCVCGYAGAATCCTGACGGNHATYTCACGGGGCGAAAGGACGCACCTGCACCCGTCGT
A  R  S  V  I  R  Q  A  R  R  I  L  T  X  I  S  R  G  E  R  T  H  L  H  P  S  F

650       660       670       680       690       700       710       720
----:----|----:----|----:----|----:----|----:----|----:----|----:----|----:----|
TTTTCACTGACTGAGGTGGTGACCCGCTTTATTTAATGATGGCGGTTGCCGTTTCAGTYCGTGGTCGCACGCCGTGATAT
F  T  D  *

730       740       750       760       770       780       790       800
----:----|----:----|----:----|----:----|----:----|----:----|----:----|----:----|
CAGGTCGCATTGATGCCGTTGCAGGGTTCCCTTGTCCATCTTCTGGAATGTTGAGACGCCGTATTGCGCTACCTGCACGG

810       820       830       840       850       860       870       880
----:----|----:----|----:----|----:----|----:----|----:----|----:----|----:----|
CGGGTGGCTGCACAACAGTCTTAATTCAAATAAGGAGATGGCAATGATGTTCACCCAACCTTTTCGTTCCGATGAGAAA
M  M  F  T  Q  P  F  R  S  D  E  K  T

890       900       910       920       930       940       950       960
----:----|----:----|----:----|----:----|----:----|----:----|----:----|----:----|
CCAARGCCCCCGAGCTTGAACTGGTGGATTTTGCATTTGGCCGCATCAAGGACGCGATTTATATCGTCAACGAAGATCAG
K  A  P  E  L  E  L  V  D  F  A  F  G  R  I  K  D  A  I  Y  I  V  N  E  D  Q

910       920       930       940       950       960       970       980
----:----|----:----|----:----|----:----|----:----|----:----|----:----|----:----|
CTGGTGGATTTTGCATTTGGCCGCATCAAGGACGCGATTTATATCGTCAACGAAGATCAGCGTTTTTGTTACGCCAATGA
L  V  D  F  A  F  G  R  I  K  D  A  I  Y  I  V  N  E  D  Q  R  F  C  Y  A  N  E

990       1000      1010      1020      1030      1040      1050      1060
----:----|----:----|----:----|----:----|----:----|----:----|----:----|----:----|
AGCAGCCAGCCAGACACTGGGCTACAGCATTACGGAATTCATGCATTTGAGCGTTGCCGATATCGATCCCGACTGGGTCG
A  A  S  Q  T  L  G  Y  S  I  T  E  F  M  H  L  S  V  A  D  I  D  P  D  W  V  A

1070      1080      1090      1100
----:----|----:----|----:----|----:----|
CTGAGCATAGCTCGCCCGACTGGTTTCAGGAATACCGCAT
E  H  S  S  P  D  W  F  Q  E  Y  R
```

Figure 1. The complete nucleotide sequence of the *hrpL* gene of *Eca* 3-2. Open reading frames for *hrpL* (b.p. 55-654) and *hrpX'* (b.p. 844-1100) are shown. Ribosome binding sites of both genes are marked bold.

These experiments have confirmed that expression of the HrpJ protein depends greatly on the growth medium composition. As expected, the HrpJ protein was barely detectable in the rich medium, more protein was produced in minimal medium, and amount of the protein was further increased by reducing nitrogen concentration in the medium from 18 to 5 mM. Additional copies of the *hrpL* gene have indeed increased the amount of HrpJ protein in *Eca* cells (Fig. 2, lanes 1-3). We were a bit surprised by the extent of this increase – cells with pTA2 grown in minimal medium (whether with normal or reduced nitrogen concentration) had 20-30 more of the HrpJ protein compared to the cells with no additional copies of the *hrpL* gene (Fig. 2, lanes 4,5). The effect of the HrpL-producing plasmid was even greater when *Eca* cells were grown in rich medium – pTA2 was effectively removing repressing effect of the rich medium on HrpJ expression (Fig. 2, lane 3 vs. lane 6). These effects could be explained to a certain extent by the fact that expression of the gene coding for the alternative sigma factor HrpL is known to depend on yet another sigma, RpoN, at least in *Pseudomonas syringae* (Hendrickson et al., 2000). We have observed that a sequence very closely resembling the RpoN-dependent promoter is located in front of the *hrpL* gene (Ageichik *et al.*, 2002). pTA2 has the sequence of this putative RpoN-dependent promoter, but *lac* promoter of the vector is located closely upstream and could drive transcription of the *hrpL* gene in rich medium.

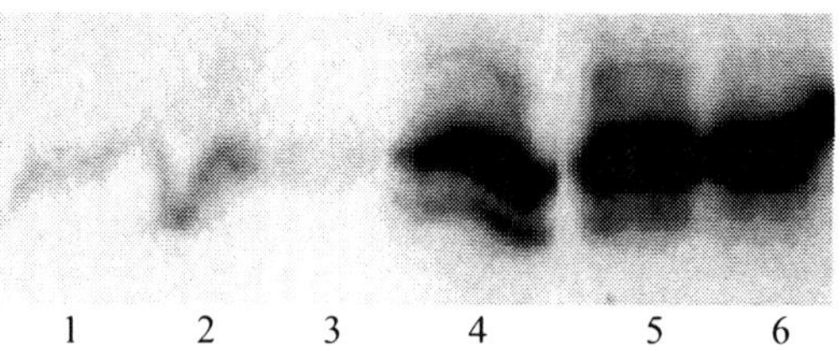

Figure 2. The effect of extra copies of the *hrpL* gene on expression of the HrpJ protein in *Eca* cells grown in different media. Strains JN42/pJQ200mp18 (lanes 1-3) and JN42/pTA3 (lanes 4-6) were grown in minimal medium A (lanes 1,3), A with nitrogen concentration reduced to 5 mM (lanes 2,4) and LB (lanes 3,6).

Increased Expression of the Type III Secretion System has Little Effect on Secretion of the HrpW-FLAG Protein

As the expression of at least one component of the TTSS was drastically increased in the presence of additional copies of the *hrpL* gene we proceeded with secretion experiments to see if increased expression of the secretory apparatus has any effect on the amount of the protein that is released into the medium. TTSS is capable of delivering its substrates to at least two different locations. Most of the so called effector proteins that usually are important virulence factors in pathogenic bacteria are translocated directly into the cells of the eukaryotic hosts. On the other hand, the second group of the TTSS substrates consists of several helper proteins secreted into the medium where they may act as secretion regulators or as components of the translocation machinery. Best studied of these proteins are harpins – glycine-rich temperature resistant hydrophobic proteins that can probably associate with plant cell walls and somehow participate in the translocation of effectors. Therefore we have chosen the HrpW harpin to compare the efficiency of the TTSS in the wild type *Eca* cells and in the cells overexpressing the TTSS. For this, we have fused the FLAG epitope to the very C-terminus of the HrpW protein and followed the distribution of this protein by western blotting with anti-FLAG monoclonal antibodies.

The HrpW-FLAG protein was secreted by the TTSS of *Eca*, although more harpin remained cell-bound. Unfortunately, induction of the TTSS by the introduction of pTA2 into the *Eca* cells with pFLAG-HrpW had little effect on secretion – the same proportion of the HrpW-FLAG protein was secreted both in the presence and in the absence of pTA2 (Fig. 3). This could be because addition of the FLAG octapeptide makes secretion of the HrpW harpin less efficient. To address this question we are now in the process of raising antibodies to the native proteins secreted through the TTSS. Another (and more likely) reason for the observed secretion inefficiency is that the TTSS, even overexpressed, still remains under control of the environmental stimuli that might be responsible for the opening of the secretion channel. Resolving this issue will require additional studies on environmental regulation of the secretion process that will be facilitated by the availability of the *Eca* strain with high level of the TTSS expression.

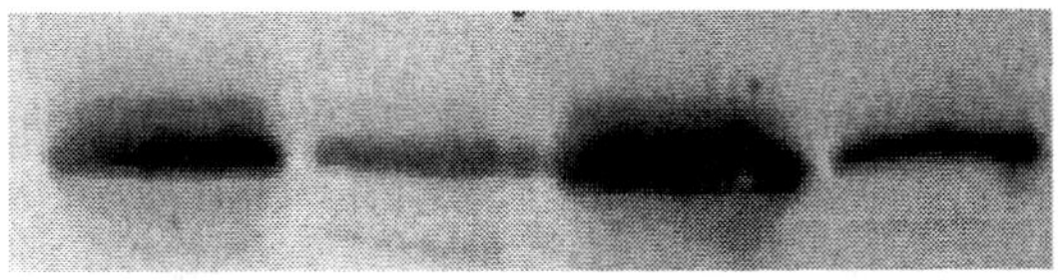

Figure 3. Secretion of the HrpW-FLAG protein by *Eca* cells. Cell-bound (lanes 1 and 3) and supernatant (lanes 2 and 4) fractions of strains JN42(pJQ200mp18, pFLAG-hrpW) (lanes 1-3) and JN42(pJQ200mp18, pTA3) grown in minimal medium A are shown.

REFERENCES

Ageichik, A.V., Evtushenkov, A.N. and Nikolaichik, Y.A.: The role of type III secretion system in *Erwinia carotovora* subsp. *atroseptica* virulence. *Plant Protection Science,* 38 (Sp. Is. 2), 523-527 (2002).

Anderson, D.M., Fouts, D.E., Collmer, A. and Schneewind, O.: Reciprocal secretion of proteins by the bacterial type III machines of plant and animal pathogens suggests universal recognition of mRNA targeting signals. *Proc Natl Acad Sci U S A*, 96 (22), 12839-12843 (1999).

Ausubel F.A., Brent R., Kingston R.E., Moore D.D., Seidman J.G., Smith J.A., Struhl K.: *Current protocols in molecular biology*. Greene Publishing Associates and Wiley-Interscience Press, New York. (1992)

Binet, R., Letoffe, S., Ghigo, J.M., Delepelaire, P. and Wandersman, C.: Protein secretion by Gram-negative bacterial ABC exporters--a review. *Gene*, 192 (1), 7-11 (1997).

Casper-Lindley, C., Dahlbeck, D., Clark, E.T. and Staskawicz, B.J.: Direct biochemical evidence for type III secretion-dependent translocation of the AvrBs2 effector protein into plant cells. *Proceedings of the National Academy of Sciences of the United States of America*, 99 (12), 8336-8341 (2002).

Chatterjee, A., Cui, Y., Chaudhuri, S., and Chatterjee, A.K.: Identification of regulators of *hrp/hop* genes of *Erwinia carotovora* ssp. *carotovora* and characterization of $HrpL_{Ecc}$ (σ^{L}_{Ecc}), an alternative sigma factor. *Molecular Plant Pathology,* 3, 359-370. (2002)

Cheng, L.W., Kay, O. and Schneewind, O.: Regulated secretion of YopN by the type III machinery of Yersinia enterocolitica. *J Bacteriol,* 183 (18), 5293-5301 (2001).

Duong, F., Lazdunski, A. and Murgier, M.: Protein secretion by heterologous bacterial ABC-transporters: The C-terminus secretion signal of the secreted protein confers high recognition specificity. *Molecular Microbiology*, 21 (3), 459-470 (1996).

Hendrickson, E.L., Guevera, P., and Ausubel, F.M.: The alternative sigma factor RpoN is required for *hrp* activity in *Pseudomonas syringae* pv. *maculicola* and acts at the level of *hrpL* transcription. *J Bacteriol,* 182, 3508-16. (2000)

Lindeberg, M., Boyd, C.M., Keen, N.T. and Collmer, A.: External loops at the C terminus of Erwinia chrysanthemi pectate lyase C are required for species-specific secretion through the out type II pathway. *Journal of Bacteriology,* 180 (6), 1431-1437 (1998).

Lloyd, S.A., Sjostrom, M., Andersson, S. and Wolf-Watz, H.: Molecular characterization of type III secretion signals via analysis of synthetic N-terminal amino acid sequences. *Molecular Microbiology,* 43 (1), 51-59 (2002).

Palacios, J.L., Zaror, I., Martinez, P., Uribe, F., Opazo, P., Socias, T., Gidekel, M. and Venegas, A.: Subset of hybrid eukaryotic proteins is exported by the type I secretion system of Erwinia chrysanthemi. *J Bacteriol,* 183 (4), 1346-1358 (2001).

Quandt, J. and Hynes, M.F.: Versatile suicide vectors which allow direct selection for gene replacement in Gram-negative bacteria. *Gene,* 127, 15-21. (1993)

Russel, M.: Macromolecular assembly and secretion across the bacterial cell envelope: type II protein secretion systems. *J Mol Biol,* 279 (3), 485-499 (1998)

Wei, Z.M. and Beer, S.V.: *hrpL* activates *Erwinia amylovora hrp* gene transcription and is a member of the ECF subfamily of sigma factors. *J Bacteriol,* **177,** 6201-10. (1995)

Xiao, Y., Heu, S., Yi, J., Lu, Y., and Hutcheson, S.W.: Identification of a putative alternate sigma factor and characterization of a multicomponent regulatory cascade controlling the expression of Pseudomonas syringae pv. syringae Pss61 hrp and hrmA genes. *J Bacteriol,* **176,** 1025-36. (1994).

SECTION 2: BIOTECHNOLOGY AND FOOD PRODUCTS

In: Biotechnology in Agriculture and the Food Industry ISBN 1-59454-119-1
Editor: G. E. Zaikov, pp. 151-160

Chapter 20

INFLUENCE OF SPIRULINA PLATENSIS BIOMASS ON MICROORGANISMS

O. B. Gorobets, L. P. Blinkova and A. P. Baturo
I.I. Mechnikov Research Institute for Vaccines and Sera (Russian Acad. Med. Sci.)
105064 Moscow, Maly Kazenny Pereulok, 5a. E-mail: instmech@iitp.ru,
Fax: +7(095) 917-5460, +7(095) 917-4900.

ABSTRACT

A blue-green microalga (cyanobacterium) Spirulina platensis and the complex of its metabolites in cultural liquid, added into culture agar at the dose of 0.01, 0.1, 1 and 10 mg/ml, can have stimulative and inhibitive influence on microorganisms. Existence of substances, which may be connected with the stimulation or inhibition of growth of some microorganisms, is supposed in Spirulina platensis biomass and in cultural liquid.

Manifestation of stimulating or inhibiting properties depended on the concentration of Spirulina and the complex of its metabolites, as well as on the way of sterilization of nutrient medium, supplemented with these substances.

Key words: Spirulina, microalga, stimulation, inhibition, microorganisms, biological activity.

In scientific publications in recent years appeared data about the influence of preparations of a blue-green microalga (cyanobacterium) Spirulina platensis (SP) on intestinal microflora [2, 3]. Apart from the stimulating influence on humans, animals and fish, an inhibiting effect with respect to some viruses and microorganisms in vivo is observed [3]. Thus, a decrease of enteropathogenic Escherichia coli and Candida albicans in human intestine is revealed. Herewith, no substantial changes in the number of lactobacilli and bifidobacteria were founded.

In our opinion, it is expedient to have information about the influence of SP as a component of bioactive additive, on microorganisms of different taxonomic groups during the experiments in vitro. That's why the aim of our research was to study the influence of SP biomass and a complex of exometabolites (cultural liquid), introduced into the composition of the culture medium, on gram-positive and gram-negative bacteria and pathogenic fungi.

Materials and Methods

In this study dehydrated biomass of SP, presented by the firm "Veta" (Russia), and lyophilised cultural liquid (CL) were used. The microalga was a semi of a commercial bioactive additive, checked for harmlessness for people. Strains of microorganisms were received from the collections of Tarasevich' Institute for Standartisation & Control and the Laboratory for microbiologic culture media. Experiments were executed on the basis of commercial nutrient agar ("Nutritious media", Mahachkala, Russia), into which we introduced aqua solutions of dehydrated SP or CL as well as commercial preparations of laminaria and fucus (as samples for comparison) at final concentrations of 0.01, 0.1, 1 and 10 mg/ml. Sterilization of microalga solutions was executed by means of boiling (regimen №1 - 100 °C 25 ± 5 min) or combination of autoclaving (120°C – 20 min) with boiling (regimen №2 - 100°C 25 ± 5 min) and also by means of chloroform (regimen №3 – 25 ± 5 min). Statistical evaluation of the data was carried out with calculation of the arithmetical mean (X) and the limit of its fluctuations (I_{95}) for sutable reliability level of date differences (p).

Results

In order to study the growth-stimulating influence of SP we made a streak inoculation of 1×10^9 CFU/ml of cultural suspense of Escherichia coli 3912/41, Salmonella typhimurium LT2, Shigella flexneri 1a NCTC 8516, Shigella sonnei "S form", Staphylococcus aureus ATCC 6538-P, Corynebacterium xerosis 1911, Proteus vulgaris R-8, Streptococcus pyogenes Dick 1, Pseudomonas aeruginosa 27/99, Candida albicans 885-653 on "hungry" 1,5% agar with or without addition of SP biomass.

The results of the experiments showed that a slight growth is observed only with C. albicans 885-653 at the concentration of SP 1 and 10 mg/ml. Slight growth in the form of «fur» is observed at the same concentrations with S. typhimurium LT2, Ps.aeruginosa 27/99 and E. coli 3912/41. Obviously, nonhydrolized SP biomass without additional nutrient sources is not enough for providing normal growth of microorganisms. The slight growth of C. albicans is probably connected with stimulating factor, which manifested itself when the potential inhibitor had no effect with respect to fungi.

On determining the absence of substantial influence of SP biomass on the growth of microorganisms, we carried out experiments for studying growth-stimulating or inhibiting effect on introducing 4 concentrations of microalgae into the nutrient agar.

The results are presented in Table 1.

Table 1. Viability of Test Straines at Different Concentrations of Spirulina Platensis Biomass

	Test strain	Regimen of sterelisation *)	Number of cells of microorganisms at different Spirulina platensis concentrations ($X \pm I_{95}$) × 10^8 CFU/ml					Authenticity of Differences (p)	
			Control	0.01 mg/ml	0.1 mg/ml	1 mg/ml	10 mg/ml	with control	between concentrations
			1	2	3	4	5		
1	Escherichia coli SU 3912/41	1	4.35±0.77 (3.58 -- 5.12)	2.87±0.56 (2.31 -- 3.43)	2.04±0.40 (1.64 -- 2.44)	1.7±0.33 (1.37 -- 2.07)	1.2±0.30 (0.9 -- 1.5)	2<0.05 3<0.05 4<0.05 5<0.05	2-3>0.05 3-4>0.05 4-5>0.05
2	Shigella dysenteriae 1362	1	1.65±0.28 (1.37 –1.93)	1.14±0.19 (0.95 –1.33)	0.145±0.024 (0.121 –0.169)	0.14±0.02 (0.12 –0.16)	0.096±0.017 (0.079 –0.113)	2≤0.05 3<0.05 4<0.05 5<0.05	2-3<0.05 3-4>0.05 4-5<0.05
3	Shigella flexneri 1a NCTC 8516	1	3.35±0.63 (2.72 –3.98)	1.8±0.27 (1.53 –2.07)	2.3±0.42 (1.88 –2.72)	3.25±0.58 (2.67 –3.83)	0.16±0.027 (0.133 –0.187)	2<0.05 3≤0.05 4>0.05 5<0.05	2-3>0.05 3-4>0.05 4-5<0.05
		2	4.98±0.74 (4.24 –5.72)	3.1±0.40 (2.7 –3.5)	1.28±0.21 (1.07 –1.49)	2.9±0.28 (2.62 –3.18)	3.61±0.44 (3.17 –4.05)	2<0.05 3<0.05 4<0.05 5<0.05	2-3<0.05 3-4≤0.05 4-5≤0.05
4	Shigella sonnei "S form"	1	3.38±0.36 (3.02 –3.74)	2.37±0.35 (2.02 –2.72)	1.82±0.30 (1.52 –2.12)	1.22±0.19 (1.03 –1.41)	0.55±0.07 (0.48 –0.62)	2<0.05 3<0.05 4<0.05 5<0.05	2-3>0.05 3-4<0.05 4-5<0.05
		2	3.44±0.58 (2.86 –4.02)	2.96±0.49 (2.47 –3.45)	2.27±0.34 (1.93 –2.61)	0.96±0.15 (0.81 –1.11)	6.71±1.14 (5.57 –7.85)	2>0.05 3<0.05 4<0.05 5<0.05	2-3>0.05 3-4<0.05 4-5<0.05

Table 1. Continued

	Test strain	Regimen of sterelisation *)	Number of cells of microorganisms at different Spirulina platensis concentrations ($X \pm I_{95}$) × 10^8 CFU/ml					Authenticity of Differences (p)	
			Control	0.01 mg/ml	0.1 mg/ml	1 mg/ml	10 mg/ml	with control	between concentrations
			1	2	3	4	5		
5	Corynebacterium xerosis 1911	1	3.0±0.54 (2.46 –3.54)	2.23±0.04 (2.19 –2.27)	0.34±0.061 (0.279 –0.401)	1.4±0.16 (1.24 –1.56)	1.7±0.21 (1.49 –1.91)	2<0.05 3<0.05 4<0.05 5<0.05	2-3<0.05 3-4<0.05 4-5>0.05
6	Salmonella typhimurium LT2	1	6.24±1.06 (5.18 –7.30)	6.2±1.05 (5.15 –7.25)	6.3±1.0 (5.30 –7.30)	9.0±1.70 (7.30 –10.70)	9.5±2.22 (7.28 –11.72)	2>0.05 3>0.05 4≤0.05 5≤0.05	2-3>0.05 3-4≤0.05 4-5>0.05
7	Streptococcus pyogenes Dick 1	1	0.0	1.72±0.27 (1.45 –1.99)	1.0±0.13 (0.87 –1.13)	1.35±0.22 (1.13 –1.57)	0.11±0.018 (0.092 –0.128)	2<0.05 3≤0.05 4<0.05 5<0.05	2-3<0.05 3-4≤0.05 4-5<0.05
8	Staphylococcus aureus ATCC 6538-P (FDA 209P)	1	6.2±1.05 (5.15 –7.25)	6.27±1.09 (5.18 –7.36)	7.78±1.32 (6.46 –9.10)	6.30±0.88 (5.42 –7.18)	1.13±0.09 (1.04 –1.22)	2>0.05 3>0.05 4>0.05 5>0.05	2-3>0.05 3-4>0.05 4-5<0.05
9	Candida albicans NCTC 885-653	1	4.4±0.66 (3.86 –5.06)	0.84±0.13 (0.71 –0.97)	0.73±0.12 (0.61 –0.85)	0.45±0.07 (0.38 –0.52)	0.50±0.075 (0.425 –0.575)	2<0.05 3<0.05 4<0.05 5<0.05	2-3>0.05 3-4<0.05 4-5>0.05

Notes: 1 – boiling at 100 °C – 25 ± 5 min.,
2 – autoclaving at 120 °C – 20 min. + boiling at 100 °C – 25 ± 5 min.

As it is seen from Table 1, the inhibiting influence of SP in comparison with the control ($p<0.05$) is revealed for all concentrations at the regimen of sterilization of culture medium by boiling (regimen №1) with respect to E. coli SU 3912/41 (max. 3.6 times), S. flexneri 1a NCTC 8516 (max. 20 times), S. sonnei «S form» (max. 6.14 times), S. dysenteriae 1362 (max. 17.2 times), C. xerosis 1911 (max. 1.76 times), C. albicans NCTC 885-653 (max. 8.8 times). Culture medium, sterilized after adding SP by means of autoclaving and further boiling (25 ± 5 min.), was used for S. sonnei «S form» strain. Here, in the highest concentration of microalga (10 mg/ml) statistically significant stimulation of microbe growth (almost 2 times) in comparison with other concentrations is revealed. Although, for S. flexneri 1a NCTC 8516 at autoclaving of the medium with microalga at the concentration of 10 mg/ml stimulation is authentically revealed only in comparison with 0,1 – 1 mg/ml doses, which had inhibitory effect in comparison with control. Sterilization of medium with microalga by boiling in some cases after maximum stimulation caused a significant decrease in viable microorganisms. Thus, growth of microalga biomass in concentration of 0,01 – 1 mg/ml was observed at growing S. pyogenes Dick 1 strain on nutrient agar with SP. This strain doesn't grow on agar without addition of stimulating factors like blood. Although, a significant repression of viability from $(1.72 \pm 0.27) \times 10^8$ CFU/ml to $(0.11 \pm 0.018) \times 10^8$ CFU/ml. As for Salmonella typhirium LT2, statistic processing of data shows that with the tendency to increase average quantities of CFU at the dose of 10 mg/ml together with growth of SP biomass concentration there is absence of significant influence SP on this microbe. Maximum size of colonies and maximum quantity of CFU at the dose of 10 mg/ml coincide with control.

The same result was obtained for S. aureus ATCC 6538-P, but here no tendency of decrease or regular increase of average quantities of survival was observed. Only the dose of 10 mg/ml caused rapid fall of staphylococcus survival to $(1.13 \pm 0.09) \times 10^8$ CFU/ml in comparison with $(6.2 \pm 1.05) \times 10^8$ CFU/ml in control, i.e. inhibition 5.5 times.

It was interesting to compare bioactivity of exoproducts of (cultural liquid) SP activity at the same concentrations in respect of analogous cultures and concentrations of the preparation. Although, drying of CL was carried out by a milder lyophilic method. The results of the experiments are presented in Table 2.

As it is seen from Table 2, in experiments with E. coli SU 3912/41 authentic stimulation of microbe growth at the doses of 0.01 – 1 mg/ml was revealed at boiling the medium with SP during 25 ± 5 min. Maximum increase of viable Esherichia coli was noted for 0.1 mg/ml dose, which was 3.5 times higher than control. Only the dose of SP contributed to a 23 times decrease of colony-forming units of E. coli SU 3912/41.

Significant growth–stimulating effect was observed for S. sonnei «S form» at autoclaving nutrient agar with 10 mg/ml of SP. Number of S. sonnei «S form» increased from $(1.45 \pm 0.22) \times 10^8$ CFU/ml to $(10 \pm 1.81) \times 10^8$ CFU/ml, which corresponded to stimulation 7,1 times. Increase of S. flexneri 1a 8516, though, at the lower degree (1.42 times) was observed at the same SP concentration and at the same regimen of autoclaving the medium with the additive.

Additive to CL medium in respect to S. pyogenes Dick 1 acted analogously to influence of SP biomass. Microbe growth was observed at addition of CL at all concentrations. Although, growth of survival lasted till the concentration of 0.1 mg/ml. Addition of 1mg/ml of CL led to the equal decrease of survival. Absence of influence of CL on the microbe is statistically confirmed for the strain of golden staphylococcus ATCC 6538-P at 0.01 – 0.1 mg/ml concentration of CL. CL concentrations 1 and 10 mg/ml caused inhibition of growth

1.46 and 6.53 times accordingly ($p < 0.05$). During this chain of experiments absence of significant influence of CL (concentrations 0.01 – 1 mg/ml) was revealed and also some increase of grown colonies of Ps. aeruginosa 27/99 (1.66 times) at the concentration of 10 mg/ml was observed.

As the thermal processing could inactivate the substances, which made conditions for stimulating (repressing) influence in SP biomass, comparative experiments were carried out. They provided sterilizing effect with chloroform and processing by means of boiling. During this chain of experiments sea algae Laminaria japonica and Fucus distichus were tested at the same concentration in culture media. The experiments were carried out with the dose of 1mg/ml, for which maximum stimulating or inhibiting effect was observed during previous experiments with SP.

The results of the experiments are presented in Table 3.

As it is seen from Table 3, processing by means of 2% chloroform during 25±5 min. In comparison with temperature processing (25±5 min., 100 °C) created authentically ($p<0.05$) milder conditions for preservation of bioactivity of SP, Laminaria and Fucus in respect of test-strains apart from C. xerosis 1911 (at addition of SP).

Viability of E. coli SU 3912/41 cells at addition of SP and Fucus, processed by chloroform, was 2 times higher; viability was 1.5 times higher when Laminaria was introduced into the medium. A higher level of CFU/ml (1.6 times) is observed in respect of C. xerosis 1911 with algae except for SP.

It is necessary to stress, that the revealed peculiarity at forming colonies of golden staphylococcus on nutrient agar with SP and Fucus was more intensive golden coloration (compared with control). S. aureus ATCC 6538-P had pale-yellow colonies at addition of Laminaria. It manifested especially visibly at processing by chloroform.

The peculiarity of algae influence, processed by chloroform, was production of colonies of a larger size (up to 4 mm). Comparison of survival level of bacteria during the experiment and control shows the same regularities, which were revealed for 1 mg/ml of SP earlier (Table 1), i.e. no tendency to increase viable staphylococcus cells was observed. Threefold stimulation of microbe survival was observed for Staphylococcus only at processing by chloroform. Slightly changed blue-green pigmentation of the additive of SP is one more proof of a milder influence of chloroform on biomass. At the same time SP biomass acquired tobacco-brown color after fulfilling the chosen regimen of autoclaving with boiling. Obviously, carotine pigments, vitamines and other bioactive substances of SP were destroyed apart from phycocyanin.

Beside intensification of bacteria growth by means of milder procession of algae additive by chloroform, the highest levels of microorganisms viability were revealed earlier (Table 1-2) after autoclaving and further boiling only at the highest concentration of SP or CL for S. sonnei «S form», S. flexneri 1a, NCTC 8516. This fact allows to assume, that the stimulator of bacteria growth, which is probably a thermosensitive substance, could not manifestate its properties, as it follows from Table 3. Naturally, even boiling without autoclaving did not lead to statistically significant growth of CFU/ml. A more probable reason of a significant intensification of bacteria growth for 10 mg/ml dose is a deep destruction of cellic structures of SP, and consequently, utilization of destructive elements as easily utilized products of nourishment.

Table 2. Viability of Test-Strains at Different Concentrations of Products of Microalga Spirulina Platensis (Cultural Liquid) Activity

	Test-strain	Regimen of influence	Number of cells of microorganisms at different CL concentrations ($X \pm I_{95}$) x 10^8 CFU/ml					Authenticity of differences	
			Control	0.01 mg/ml	0.1mg/ml	1 mg/ml	10 mg/ml	with control	between concentrations
			1	2	3	4	5		
1	E. coli SU 3912/41	1	2.30 ± 0.37 (1.93-2.67)	6.0±1.2 (4.8-7.2)	8.0±1.4 (6.6-9.4)	4.5 ± 1.25 (3.25-5.75)	0.1±0.015 (0.085-0.11)	2<0.05 3<0.05 4<0.05 5<0.05	2-3>0.05 3-4>0.05 4-5>0.05
2	S. flexneri 1a NCTC 8516	1	4.45± 0.75 (3.70-5.20)	3.25±0.33 (2.92-3.58)	2.96±0.50 (2.46-3.46)	0.94±0.15 (0.79-1.09)	0.12±0.013 0.107-0.133)	2≤0.05 3≤0.05 4<0.05 5<0.05	2-3>0.05 3-4>0.05 4-5>0.05
		2	5.0± 0.75 (3.70-5.20)	2.20±0.30 (1.90-2.50)	0.58±0.09 (0.49-0.67)	0.087± 0.014 (0.073-0.101)	7.10±1.13 (5.97-8.23)	2<0.05 3<0.05 4<0.05 5≤0.05	2-3<0.05 3-4<0.05 4-5<0.05
3	S. sonnei «S form»	1	1.45± 0.22 (1.23-1.67)	1.43±0.24 (1.19-1.67)	0.9±0.14 0.76-1.04)	0.6± 0.09 (0.51-0.69)	0.51±0.091 (0.419-0.601)	2<0.05 3<0.05 4<0.05 5<0.05	2-3<0.05 3-4<0.05 4-5≤0.05
		2	1.4± 0.21 (1.19-1.61)	1.4± 0.23 (1.17-1.63)	1.34± 0.21 (1.13-1.55)	1.24± 0.20 (1.04-1.44)	10± 1.81 (8.19-11.81)	2>0.05 3>0.05 4>0.05 5>0.05	2-3>0.05 3-4>0.05 4-5<0.05

Table 2. Continued

	Test-strain	Regimen of influence	Number of cells of microorganisms at different CL concentrations $(X \pm I_{95}) \times 10^8$ CFU/ml					Authenticity of differences	
			Control	0.01 mg/ml	0.1mg/ml	1 mg/ml	10 mg/ml	with control	between concentrations
			1	2	3	4	5		
4	P. aeruginosa 27/99	1	1.8± 0.23 (1.57-2.03)	1.88± 0.28 (1.60-2.16)	1.97± 0.031 (1.66-2.28)	2.0± 0.26 (1.74-2.26)	3.0± 0.51 (2.49-3.51)	2>0.05 3>0.05 4>0.05 5<0.05	2-3>0.05 3-4>0.05 4-5<0.05
5	S. pyogenes Dick 1	1	0.0 (no growth)	0.86± 0.11 (0.75-0.97)	1.06± 0.18 (0.88-1.24)	0.1± 0.017 (0.083-0.117)	0.10± 0.0382 (0.0618-0.1382)	2<0.05 3<0.05 4<0.05 5<0.05	2-3>0.05 3-4<0.05 4-5>0.05
6	S. aureus ATCC 6538-P (FDA 209P)	1	6.27± 1.0 (5.27-7.27)	6.45± 1.67 (4.78-8.12)	7.26± 1.3 (5.96-8.56)	4.3± 0.77 (3.67-0.07)	0.96± 0.13 (0.83-1.09)	2>0.05 3>0.05 4≤ 0.05 5<0.05	2-3>0.05 3-4<0.05 4-5<0.05

Note: 1 – boiling at 100 °C – 25 ±5 min.,
2 – autoclaving at 120 °C – 20 min. + boiling at 100 °C - 25 ± 5 min.

Table 3. Influence of Thermal Processing and Chloroform on Bioactivity of Algae in Culture Media

Regimen of sterilisation	Algae	Number of cells after 19±1 hr ($X \pm I_{95}$) x 10^8 CFU/ml					
		E.coli SU 3912/41		C.xerosis 1911		S. aureus ATCC 6538-P	
		Experiment	Control	Experiment	Control	Experiment	Control
100° C 25±5 min.	Spirulina platensis	4.0 ± 0.72 (3.28 -- - 4.72)		1.13 ± 0.19 (0.94 -- 1.32)		1.66 ± 0.29 (1.37 -- 1.95)	
	Laminaria japonica	2.77 ± 0.12 (2.65 -- 2.89)	7.4 ± 1.33 (6.07 -- 8.73)	0.42 ± 0.07 (0.35 -- 0.49)	1.46 ± 0.26 (1.2 -- - 1.72)	0.90 ±0.15 (0.75 -- - 1.05)	1.91± 0.34 (1.57 -- 2.25)
	Fucus distichus	1.54 ± 0.26 (1.28 -- 1.80)		0.30± 0.048 (0.252 -- 0.348)		0.60 ± 0.10 (0.50 -- - 0.70)	
Chloro-form 2 % 25±5 min.	Spirulina platensis	8.65 ± 1.55 (7.1 -- - 10.2)		1.15 ± 0.18 (0.97 -- 1.33)		5.0 ± 0.9 (4.1 -- 5.9)	
	Laminaria japonica	4.1 ± 0.73 (3.37 -- 4.83)	7.4 ± 1.33 (6.07 -- 8.73)	0.71 ±0.12 (0.59 -- 0.83)	1.46 ± 0.26 (1.2 -- - 1,.2)	1.94 ± 0.32 (1.62 – 2.26)	1.91 ± 0.34 (1.57 -- 2.25)
	Fucus distichus	3.16 ± 0.53 (2.63 -- 3.69)		0.50 ± 0.08 (0.42 -- 0.58)		1.0 ± 0.17 0.83 -- - 1.17)	

Discussion

It is common knowledge that SP has a high nutrient value, connected with existence of 55-70 % albumen, up to 15-25% carbohydrate, to 5% lipids, 4% nucleic acids, 7% mineral substances in SP cells. SP is considered a perspective substrate as a basis of cultural media [2, 4] thanks to a vast range of vitamins, such as leucine, valine, isoleucine, threonine, glutamine and asparcamic acids, alanine, arginine, irreplaceable fatty acids (gamma-linoleic and palmitic acids predominate). In this respect it is possible to propose, that a certain number of destroyed cellic substrate (e.g. after autoclaving), which suits for utilization by some types of microorganisms, can be found at high concentrations of SP additives. But existence of inhibiting substances in SP products can interferate with growth-stimulating factors of microalga. Inhibiting influence of SP on C. albicans or E. coli, revealed during experiments in vitro, agrees with literary information about the repressing effect of SP on these microorganisms in enteric conditions. Information about the absence of inhibiting influence of SP on bifidobacteria and lactic acid bacilli [3] is also compared with the results of our experiments, from which it follows, that some types of bacteria (P. aeruginosa 27/99 or S. aureus ATCC 6538-P) do not undergo the influence of microalga at doses of 0.01-1 mg/ml.

From practical positions the fact, that SP can provide growth of a fastidious microbe – hemolitic streptococcus S. pyogenes Dick 1 without addition of such ingredients as blood, serum, or yeast.

Thus, within the culture medium (artificial medium of inhabitance) of SP and its exometabolites can produce both stimulating and inhibiting influence on microorganisms. This influence in some cases is similar to the influence of alga on them in microecological conditions of their natural medium of inhabitance.

References

[1] I.P. Ashmarin, A.A. Vorobiov: *Statistical methods in microbiological researches*. Medgiz, Leningrad, 1962, 180 p. (in Russian).

[2] L.P. Blinkova: Biological activity of Spirulina. *Zh. Microbiol.,* Moscow (2), 114 (2001).

[3] R. Kozlenko, R.H. Henson: The study of Spirulina: Effects on the AIDS Virus, Cancer and the Immune System. *Healthy & Natural Journal*, (3), 5 (1997).

[4] *Earth food Spirulina: A new food to restore our health and planet*. Spirulina Health Library. 1994, July.

In: Biotechnology in Agriculture and the Food Industry ISBN 1-59454-119-1
Editor: G. E. Zaikov, pp. 161-164

Chapter 21

HYGIENIC SAFETY OF RAW MILK

***I. M. Burykina*[1] *and S. V. Molotov*[2]**

[1]Vologda State Dairy Farming Academy named after N.V.Vereshagin, Vologda, Molochnoe, Shmidt str., 2

E-mail: burikina @visp.ru, fax (8172) 76-16-85

[2] Experimental Dairy Plant of Vologda State Dairy farming Academy named after N.V.Vereshagin, Vologda, Molochnoe, Pankratov str., 15

E-mail: uomz@ vologda.ru, fax (8172) 76-16-02

ABSTRACT

Raw milk contains specific and non-specific microflora, which influences the development of biotechnological processes during lactic souring.

The quality of raw milk is determined by the conditions of its production and primary milk processing on farm.

The quantitative structure of microflora changed depending on conditions of primary milk processing on farm and especially of conditions of delivery of milk to the plant; and also oscillated on seasons of year and has compounded:

- heat-resistant micro-organisms –4,20 ±0,03 lg/cm^3
- Staph.aures 4,12 ±0,02 lg/cm^3;
- Enterococcus bacteria - $1 \cdot 10^3$ -$2 \cdot 10^5$ K.G.U./ cm^3
- psychrotrophus microorganisms –$8,6 \cdot 10^3$ - $9,2 \cdot 10^7$ K.G.U./cm^3

During researches it is revealed, that as sources of chemical danger act equipment and stock manufactured of materials, not allowed for a contact to foodstuff; forage inclusive antibiotics; residual quantities of detergents and disinfectants.

Keywords: Quality of raw milk, Microorganisms, Testing, Chemical and physical changes of products, Control of sanitary laboratory, Microbiological impurity

In process of complication of economical and ecological conditions in the country the quality of food products for last years has deteriorated on many indexes.

Raw milk contains as specific and non-specific microflora, which influences on the development of biotechnological processes during lactic souring.

The speciality of milk processing branch, as however of other food manufacturing industry branches is connected with the fact that the quality of used raw directly connected with formation of safety indexes of end product.

Presence of psychotrophe microorganisms causes rancid, bitter or putrefactive flavour.

One of the most urgent state problems is to supply population with quality food. In process of complication of economical and ecological conditions in the country the quality of food products for last years has deteriorated on many indexes.

Along with a decrease of technological properties and nutritional value there is an impairment of ecological safety of food products. Under the data of State Sanitary-Epidemiological Supervision, as a whole on Russia for 1996-2001 years the specific gravity weight of samples of food raw and foodstuff which are not adequate the standards on sanitary-chemical parameters makes 4,8 – 5,9 %, on sanitary-microbiological parameters 6,9 – 7,4 %. Including on milk and milk products makes accordingly 3,5-5.9 % on chemical parameters and 9,1 – 10,6 % on microbiological parameters. The unfavorable situation is determined also by the fact that bromatoxisms do not decrease. Under the data of State Sanitary- Epidemiological Supervision per separate years the number of mass bromatoxisms reached 50 - 70 per one year, the number of damaged people has compounded 1600 men.

In recent years such little-known infections as iersinious, listerosis, campilobacterios, different arboviral and ricketsial fevers etc. are more often registered. On nature and levels of risk of foodstuff for man health on the first place are impurities of microbiological origin. The activity of any micro-organisms provokes chemical and physical changes of products. More often these processes are unwanted and result in change of concistence, pH, taste, and sometimes in damage of products. A special attention is merited by micro-organisms, causing bromatoxisms.

According to a list of foodstuff on a rate of fouling by microorganisms and frequency of cases of bromatoxisms elaborated by World-wide organization of public health services, milk and the dairy products are referred to the first category (foodstuff of their components, which most frequently serve the direct source of bromatoxisms).

The speciality of milk processing branch, as however of other food manufacturing industry branches is connected with the fact that the quality of used raw directly connected with formation of safety indexes of end product.

The quality of milk – raw material is determined by the conditions of its production and primary milk processing on farm. Milk even at obtaining in good sanitarian conditions is not a sterile product. Raw milk contains as specific and non-specific microflora, which influences on the development of biotechnological processes during lactic souring. The number of bacteria in milk, obtained from healthy cows is insignificant - from 1000 up to 10 000 in 1 ml.

At disturbance of sanitarian and technological rules (milking, primary milk processing, stock and equipment washing) nonspecific microflora is met in milk: putrescence bacteria, butryc acid bacteria, coliform bacteria, mold fungi, as well as morbific microorganisms (tuberculosis, brucelosis, desinteria pathogenic organisms and others).

Pathogenic microorganisms can fall into milk from ill animals, sick men and carriers of bacilli.

Hygienic properties of milk depend on many factors. The sources of biological danger of milk-raw are: water, air, condition of health of animals and men maintaining farms.

Control of hygienic indexes is conducted on one side by milk industry enterprises and on the other by supervision of sanitary service. Laboratory control data of Vologda Standard Center for 2002 testify:

- -toxic elements, microtoxins, pesticides are not revealed in raw milk;
- -pathogenic and morbific micro-organisms are not revealed in raw milk;
- -in milk from Vologda district farms antibiotic of tetracycline and penicillin group are revealed, specific gravity of unsatisfactory probes is equal to 0,5 %.

The researches of quality of milk, received from Vologda, Kirillov, Tcherepovets, Griazovets districts Vologda region to The Experimental Dairy Plant of the Vologda State Farming Academy on microbiological parameters have shown:

- raw milk contains from 4,8·104 up to 1,8 ·108 cells of microorganisms in 1 cm^3;
- up to 50% from general microbial sowing make heat-resistant micro-organisms,
- contents of somatic cells 105 –107 in 1 cm^3.

The researches have shown, a reason of microbiological impurity of milk is the non-observance of terms of surrender of milk on milk processing firm from animals, subjected to treatment by antibiotics; violation of modes of washing both disinfection of the milking equipment and equipment for primary processing of milk.

The quantitative structure of microflora changed depending on conditions of primary milk processing on farm and especially of conditions of delivery of milk to the plant; and also oscillated on seasons of year and has compounded:

- heat-resistant micro-organisms –4,20 ±0,03 lg/cm^3
- Staph.aures 4,12 ±0,02 lg/cm^3;
- Enterococcus bacteria - 1·10^3 -2·105 K.G.U./ cm^3
- psychrotrophus micro-organisms –8,6 ·10^3 - 9,2 ·107 K.G.U./cm^3
- osmotolerant micro-organisms –1,1 ·10^3 - 7,3 ·104 K.G.U/cm^3

The detected micro-organisms are capable to call defects of organoleptic parameters of milk and dairy products, including taste and consistence. Aerobic bacillus B.cereus, B mycoides, B.polymyxa make a clot in milk at a low acidity.At development these micro-organisms excrete exoferments under effect of which clot with bitter taste will be derivated.

M.varians, M. caseoliticum give positive reacting to a coagulase and coagulate milk at a low acidity, decomposing milk proteins.

The development of M. lutens results to acidification of milk without curdling.

B.subtilis, L.lactis subsp. maltigenes in summer period can call changes of taste, creating malt aroma and flavour.

Presence of psychotrophe micro-organisms causes rancid, bitter or putrefactive flavour.

The quality of used raw is connected to formation of safety parameters of finished product.

For a decrease of bacterial sowing of raw milk observance of the sanitary-hygienic requirements at all stages is very important:

- care of agricultural animals(maintenance of purity of dermal cover, especially udder skin, back extremities, tail, observance of rules of premilking preparement, high quality of forages).
- obtaining of milk from healthy animals (regular veterinary examinations observance of rules of usage of milk from ill animals).
- maintenance of purity of milking machines, milk ware and stock (observance of periodicity both rules of washing and desinfection).
- health of the serving staff (periodic medical examinations and hygienic training)
- primary processing of milk (filtration, immediate cooling, observance of the terms of storage)
- transportation of milk (specialized transport, time of delivery)

During researches it is revealed, that as sources of chemical danger act equipment and stock manufactured of materials, not allowed for a contact to foodstuff; forage inclusive antibiotics; residual quantities of detergents and disinfectants

An important phase of activity of any milk plant should be close relationships of the specialists of milk plants and agricultural firms.

The experience of Vologda Dairy Combine and Experimental Dairy plant demonstrates, that constant sanitary and educational work as well as methodical help gives the workers of milk-commodity farms positive results in obtaining qualitative raw.

REFERENCES

[1] Safety of Russia. *Legal, social-economic and scientific-technical aspects. Food safety.* Part 2. Znanie, Moskaw, 2001. 480 p. (in Russian).

In: Biotechnology in Agriculture and the Food Industry ISBN 1-59454-119-1
Editor: G. E. Zaikov, pp. 165-170

Chapter 22

COMPLEX USE OF *PYTHIUM DEBARYANUM* BIOMASS – A PRODUCER OF PUFA

***V. I. Dukova*[1], *E. P. Tkachevskaya*[1], *I. V. Konova*[2], *L. A. Galanina*[2], *Ya. E. Sergeeva*[2], *E. A. Alyab'ieva*[1], *E. E. Joukova*[1], *A. F. Mironov*[1] and *R. P. Evstigneeva*[1]**

[1]The Lomonosov State Academy of Fine Chemical Technology.
119571, Moscow, pr. Vernadskogo, 86
E-mail:httos.mitht@g23.relkom.ru
Fax: +7 095 434 8678
[2] Institute of Microbiology, Russian Academy of Sciences
Prospect 60-Letiya Octyabrya 7, korp. 2, Moscow 117312, Russia
E-mail: ina@host.inmi.ru
Fax: +7 095 135 6530

ABSTRACT

The biomass of oomycete *Pythium debaryanum* (strain INMI) has been prepared by the method of deep cultivation and proposed to be used for the enzyme preparation production (specific lipoxygenase (LOX) activity 1.5 ± 0.3 μM hydroperoxylinoleate/(min·mg protein)) and lipids with high content of polyunsaturated fatty acids (PUFA) (up to 56 % of total fatty acids).

Step-by-step methods of enzyme sedimentation by ammonium sulfate and gel-chromatography were used for the production enzyme preparation with LOX activity. LOX preparation was characterized by specific LOX activity, spectrophotometric and chromatographic methods. Lipids had been extracted from biomass after LOX isolation (yield 8±0.1% of dry biomass). GLC analysis of fatty acids composition of the total lipids, which was obtained from native biomass and from biomass after LOX isolation, has not revealed any considerable differences in the contents of PUFA.

Thus, biomass of oomycete *Pythium debaryanum* may be considered not only as a source of pharmacologically active lipids, but also as a source of enzyme preparation with LOX activity.

Key words: Oomycete, *Pythium debaryanum*, lipoxygenase, lipids, PUFA.

Microorganisms are seldom considered as a source of lipoxygenases. Oomycetes of the genus *Pythium*, the culture *Pythium debaryanum* in particular, can be considered as a producer of polyunsaturated fatty acids (PUFA) including eicosapolyenoic fatty acids (EFAs), dominated by arachidonic ($C_{20:4}$) and eicosapentaenoic ($C_{20:5}$) acids. These PUFA can serve as substrates for lipoxygenases (LOX). The purpose of the present work was the studying of opportunity of complex use of *Pythium debaryanum* biomass for the extraction of pharmacologicaliy active lipids with high content PUFA and isolation of enzyme preparation with lypoxygenase activity.

Oomycetes of the family *Pythaceae* are widespread, inhabit in soil, and also parasitize on plants. Oomycetes of the genus *Pythium* are of interest from the point of view of their ecologic-geographical spreading and biological features. *Pythium debaryanum* is phytopathogen, which causes an infection of some plants. [1]. This oomycete is an active producer of PUFA. It was considered, that the fractions of lipids, including active component as polyunsaturated fatty acids (arachidonic and eicosapentaenoic), are responsible for the pathogenicity of oomycetes from genera *Phytophthora* and *Pythium* [2].

The strain *P.debaryanum* was selected and investigated in Institute of Microbiology Russian Academy of Sciences as an active producer of EFAs. The physiological features of its growth and lipogenesis are revealed. Also the bases for biotechnological production of lipids, containing C_{20}-polyens, synthesized by this oomycete are developed [3, 4].

When lipids of *P.debaryanum* were synthesized *de novo* in favorable conditions of growth oomycete on glucose-mineral medium with bactopeptone, the unsaturated fatty acids prevailed (up to 78%) in the sum of total fatty acids, namely oleic acid ($C_{18:1}$) (up to 23%), EPAs (up to 33 %, mainly arachidonic and eicosapentoenoic acids), C_{18}-poleyns (linoleic acid ($C_{18:2}$) and linolenic acid ($C_{18:3}$) acid up to 18.5 and 4.5 % accordingly) and in minor fractions palmitoleic acid ($C_{16:1}$). The predominant saturated fatty acid was palmitic acid ($C_{16:0}$) (up to 17 %), other saturated acids (such as stearinic ($C_{18:0}$), myristic ($C_{14:0}$), lauric ($C_{12:0}$), arachidic ($C_{20:0}$)) were presented in minor or trace amounts.

The pharmacological activity of the lipids *P.debaryanum* was tested. Pre-clinical experiments performed at Sechenov's Moscow Medical Academy have shown that thin watery emulsion obtained on the basis of oomycete lipids reduces the content of cholesyerol [5].

Thus, it is possible to consider investigated strain *P.debaryanum* as perspective producer of pharmacologically active lipids.

PUFA (including linoleic acid and arachidonic acid) are substrates for lipoxygenase. We have tested lipoxygenase activity of the enzyme preparation isolated from oomycete *P.debaryanum* and have developed a methods of complex use of biomass of oomycete in order to produced both lipids and enzyme preparation.

Lipoxygenase (linoleate:oxygen oxidoreductase, EC 1.13.11.12) is an enzyme which catalyzes the oxidation of unsaturated fatty acid containing the *cis, cis*–1,4–pentadiene unit (such as linoleic, linolenic and arachidonic acids) by molecular oxygen and produces conjugated *cis, trans*-hydroperoxides of these fatty acids. The products of PUFAs enzymic oxidation and its metabolites in plants and animals cells (jasmonic acid, oxilipins, eicosanoids etc.) have a wide spectrum of biological activity, and take part in regulation of many metabolic processes [7].

LOX is widely distributed in plants, animals and have been isolated from various plants and animals tissues and characterized, whereas microbial lipoxygenases are seldom investigated. The discovery of a manganese-containing lipoxygenase (Mn-LOX) from the fungus *Gäumannomyces graminis*, and its unique ability to oxygenate on the fatty acid carbon between two *cis*-double bonds (C_{11} in linoleic acid) provides new prospects in studying of microbial lipoxygenases. [6]. LOX activity was detected only for a few oomycetes: *Lagenidium giganteum* [8], *Saprolegnia parasitica* [9], *Achlya ambisexualis* [10].

The investigation of producers of biologically active substances, their lipids and LOX activity are very important for medicine, food industry and bioorganic chemistry

In our work, enzyme preparation was isolated from oomycete *P.debaryanum* by precipitation with ammonium sulfate. The enzyme preparation was purified from low-molecular impurity by gel-chromatography on sephadex. In the present work, two ways of isolation of enzyme preparation (homogenization with quartz sand and biomass destruction by ultrasound on ice) were compared. It was found that the homogenization by ultrasound on ice resulted in increasing of the protein content 4 times and increasing the lipoxygenase activity 1.5 times (in comparison with homogenization with quartz). From 100 g of biomass (H_2O contents about 80 %) was isolated 40 ml of LOX preparation with the protein contents 1.8±0.05 mg/ml. The isolated enzyme preparation contained no more than 10 % of impurity according to HPLC data. A 100 kDa molecular weight of purified LOX was determined by SDS gel-electrophoresis on PAAG.

The methods of defining lipoxygenase activity are well developed for enzyme preparations from animals and plants sources. The lipoxygenase activity of enzyme preparation has been confirmed by interaction with lipids substrates (linoleic acid, linolenic acid, arachidonic acid). [11]. In our experiments, linoleic acid ($C_{18:2}$) was used as lipids substrate. The lipoxygenase activity was detected by spectrophotometrical and polarographical methods [12].

Specific LOX activity of the isolated preparation was 1.5±0.3 μM hydroperoxylinoleate/(min·mg of protein). The oxygen absorption was 0.2 ± 0.005 O_2 μM/min during enzymatic oxidation of linolenic acid by LOX *P.debaryanum*.

We have found the optimal conditions for enzymatic oxidation linolenic acid by LOX *P.debaryanum*: pH 9.3 (the tetraborate buffer), concentration of protein 15÷20 μg/ml, the concentration of linolenic acid $2.8\times10^{-6} \div 3.3\times10^{-6}$ M. The preparation of LOX *P.debaryanum* was thermostable, the speed of products formation practically did not change by the temperature interval from 15 °C up to 30 °C.

Also the chromatographic analysis of products of enzymatic oxidation linoleic acid by LOX *P.debaryanum* was carried out. The HPLC results have indicated, that the contents of hydroperoxides of linoleic acid in reaction products achieved 88.5%.

The enzymatic activity of *P.debaryanum* preparation was not inhibited by indometacin (10^{-4}M) – inhibitor of cyclooxygenases, and was strongly inhibited by nordihydroguaiaretic acid (10^{-5}M). The absence of optical absorption in an electronic spectrum of enzyme preparation at 370-460 nm suggested the absence of heme-containig impurity. Thus, the presence of lipoxygenase activity in enzyme preparation of *P.debaryanum* was confirmed.

In the definition of lipoxygenase activity of *P.debaryanum* we used biomass of the oomycete cultivated in media with two different types of nitrogen sources. In the first variant (variant A) on glucose-mineral media with bactopeptone the lipids were synthesized *de novo*.

The assimilation and probably conversion of exogenous fatty acids occurred in the second variant (variant B) where bactopeptone was replaced by soybean flour. The content PUFA was practically identical: accordingly 42.4% and 46.9 % in the sum of total fatty acids in biomass of oomycete grown in A and B types of media. However the content sum of EPAs was 25.4 % at synthesis *de novo* (variant A) and 14.1 % at presence of exogenous C_{18}-polyenes contained in soybean flour (variant B). It should be noted that in variant B the content of linoleic acid (up to 30.0 %) was about 2 times higher than at synthesis *de novo* (variant A).

The definition LOX activity of preparations (μM of hydroperoxide/min·mg of protein), isolated from biomass grown in the two compared media (variants A, B) has shown, that this enzymatic characteristic was 1.8 and 1.5 accordingly.

The total lipids were extracted with chloroform - methanol mixture from biomass of *P.debaryanum*. The GLC analysis of fatty acids methyl esters (FAMEs) of total lipids, extracted from biomass after lipoxygenase isolation, has not revealed considerable differences in the PUFA contents in comparison with the same, extracted from native biomass (42.6±0,1 % and 41.9±0,1 % respectively respectively).

It is possible to raise the content of PUFA by method of urea inclusion adducts and in our case the PUFA content has been raised up to 56%.

The results of our work have shown that the investigated oomycete *P.debaryanum* (strain INMI) can be used not only as a source of pharmacologically active lipids with the high PUFA content, but also for production of enzyme preparation with lipoxygenase activity.

EXPERIMENTAL PART

1. Organism and Cultivation Conditions

The oomycete *P.debaryanum* (strain INMI), selected in the Institute of Microbiology Russian Academy of Science, was supported on a wort agar medium. Cultivation (method deep growth) was performed as described earlier [4, 12].

2. Extraction and Analytical Procedures

Lipidologocal methods (extraction of lipids, acidic methanolysis of total lipids) were taken from hahdbook [13]. The amount of extracted lipids was measured gravimetrically, as a percentage of the dry biomass. Fatty acid methyl esters (FAMEs) were analysed by gas-liquid chromatography (GLC) on model 3700 chromatograph ("Chromatograph", Russia) equipped with a flame-ionisation detector and 2500x3 mm glass column packed with 17% diethylenglycol succinate on Chomosorb WAW-DMCS-HP, 80-100 mesh (Lachema, Czech Republic). The carrier gas was argon at a flow rate of 50 ml/m. The temperature of the column and evaporator were 180 °C and 250 °C respectively. FAMEs were identified by comparison of their retention times with those of the standard samples and quantified by the triangular method [13].

3. Isolation of Lipoxygenase Preparation from Biomass

LOX preparation was obtained from cells of oomycete *P.debaryanum*, biomass in PBS buffer (0.01M, pH 7.2, 0.9 % NaCl) was homogenized by ultrasound on ice 15 min or homogenized with quartz sand. After isolation of biomass, filtrate was processed consistently by ammonium sulfate solution (30 % and 60 % from saturation). After centrifugation (15 min at 14 000 rpm) the precipitate contained enzyme preparation of *P.debaryanum* lipoxygenase. The enzyme preparation was purified by gel-filtration on sephadex G-25 and G-100.

Enzyme prparation was analysed by HPLC on chromatograph "Beckman-System Gold" (USA) equipped with the UV-detector (λ 280 nm), column 250x2,2 mm, C-18, 5 μm, "Ultrasphere ODS". Eluent flow rate was 0.25 ml/min with gradient mode. The mobile phase A: 0.1 % a solution of trifluoroacetic acid in water; B: acetonitrile and 0.07 % of trifluoroacetic acid; a gradient 0%-90% B. The content of separable components was estimated by method of internal normalization of the peaks areas.

Concentration of protein in enzyme preparation was determined according to Bredford's method [14].

The dependence of enzyme activity on concentration of lipids substrate (linoleic acid, Sigma (USA)), on concentration of enzyme (protein) and on pH of medium were investigated by spectrophotometric method in which the accumulation of hydroperoxylinoleate was detected by growth of optical absorption at 234 nm. LOX activity detected as concentration of linoleic acid hydroperoxides formed during enzymatic reaction (λ_{234}, $\varepsilon = 27\,400\ M^{-1}\ sm^{-1}$) i.e., μmole hydroperoxylinoleate/(min·mg of protein). The experiments were carried out in microcuvette of spectrophotometer " Jasco UV 7800 series " (Japan).

For enzymatic oxidation of linoleic acid solution of linoleic acid in ethanol 0,75 ml (3.3×10^{-6} M) was added to solution of protein 25 ml (20 mkg/ml) in buffer pH (9.3). The enzymatic reaction was controlled by thin–layer chromatography on Silufol in system: hexane - diethyl ether - acetic acid, (8:7:0.1), using vanillin or/and iodine (R_f of hydroperox. = 0.35, R_f of acids = 0.55). The reaction was stopped by acidification 5 % HCl up to pH 4-5. The products of the reaction were isolated by extraction with diethyl ether (3x30 ml).

The products of enzymatic reaction were fractionated by using reverse-phase chromatography on chromatograph Gilson UV/VIS - 155 (column is reverse-phase ULTRASPERE BECKMAN (USA), ODS, 5 μ 4.6 mm x 25 cm, equipped with UV-detector with the filter on 234 nm and 245 nm), eluent acetonitrile-water, 70:30. The detection of the divided component contents (hydroxy -, hydroperoxy -, keto- derivatives of linoleic acid, linoleic acid) was carried out by the method of internal normalization of the areas of peaks in comparison with the standards.

References

[1] N. S. Novotel'nova, K. A. Pystina, O. A. Golubeva: *The peronosporales fungi – the pathogen of cultural plants in USSR*. Nauka, Leningrad, 1979. 152 p. (in Russian).

[2] O. A. OzeretckovskayA: The induction of plants stability by biogenic elysitors of phytopathogens. *Applied biochemistry and microbiology* (Rus.), **30** (3), 325 (1994). (in Russian).

[3] N.L. Solov'eva, I. V. Konova, L. A. Galanina, N. K. Babanova: Lipids of *Pythium debaryanum*. *Microbiology* (Rus.), **66** (2), 475 (1997). (in Russian).

[4] L. A. Galanina, N. L. Solov'eva, I. V. Konova: Synthesis of eicosapolyenoic acids in *Pythium debaryanum* grown in various media. *Microbiology* (Rus.), **68** (4), 473 (1999). (in Russian).

[5] N. L. Solov'eva, N. K. Babanova, I. V. Konova, L. A. Galanina, V. V. Kaptcov: Obtaining of a liposomal medicinal preparation involving eicosapentaenoic acid-containing lipids. III *Russian scientific congresses «Humans and Medicines»*. Farmmedinfo, Moscow, 1996, p. 49. (in Russian).

[6] S. Chao, E. Oliv: Manganase lipoxygenase. Purification and characterization. *J. Biolog. Chem.*, **273** (21), 13072 (1998).

[7] R. Brash: Lipoxygenases: Occurrence, Functions, Catalysis, and Acquisition of Substrate. *J.Biol. Chem.*, **274** (34), 23679 (1999).

[8] P. K. Stumpf et al. (Eds): *The metabolism, structure and function of plant lipids*. Plenum, New York, 1987. p. 421.

[9] R. P. Herman, M. Hamberg: Properties of the soluble arachidonic acid 15-lipoxygenase and 15-hydroperoxide isomerase from the oomycete *Saprolegnia parasitica*. *Prostaglandins*, **34** (1), 129 (1987).

[10] R. P. Herman, M. M. Luchini, C. A. Herman: *Exp. Mycol.*, **13** (1), 95 (1989).

[11] C. Benedetto et al. (Eds): *Prostaglandins and related substances, a practical approach*. IRL press Oxford, 1987 p.229.

[12] Dukova V. I., Demcheva E. A., Tkachevskaya E. P., Sergeeva Ya. E., Galanina L. A., Konova I. V., Evstigneeva R. P.: The influence of vitamin K_1 on lipoxygenase activity oomycete *Pythium debaryanum* – producer of PUFA. *Biotechnology* (Rus.), **6** (in press) (2003). (in Russian).

[13] M. Kates: *Tekhnika Lipidologii*. Mir, Moscow, 1975. 322 p. (in Russian).

[14] R. M. C. Dawson, D. C. Elliott, W. H. Elliott, K. M. Jones: *Spravochnic biokhimika*. Mir, Moscow, 1991. 543 p. (in Russian).

In: Biotechnology in Agriculture and the Food Industry
Editor: G. E. Zaikov, pp. 171-173
ISBN 1-59454-119-1

Chapter 23

RECEPTION OF LACTULOSE PREPARATIONS FOR PRODUCTS OF FUNCTIONAL NUTRITION

A. G. Hramtsov, I. A. Evdokimov, S. A. Ryabtseva, A. D. Lodygin and L. R. Alieva
North-Caucasian state technical university, Stavropol
355029, Kulakov av., 2, Stavropol, Russia,
e-mail:, hramtsov@stv.runnet.ru, fax (8652) 94-59-10.

ABSTRACT

Lactulose is carbohydrates with brightly expressed prebiotic by properties and is recognized all over the world as the bifidogenic factor № 1. In clause the theoretical bases and ways of lactulose reception are considered. The results of own researches of the authors on development of technology of lactulose preparations with use of baromembrane and electromembrane treatment of dairy raw material are submitted. The analysis of directions of lactulose use is carried out by manufacture of functional products of a nutrition.

Key words: bifidogenic factors; lactulose; isomerization; baromembrane and electromembrane treatment; functional products

In the XXI century in all developed countries the problem of recovery and maintenance of normal intestinal microflora, which plays a huge role in the human health, is an essential part in state policy. Probiotics, prebiotics and sinbiotics are applied with this purpose. A lot of substances with different nature, structure and properties, are considered as prebiotics or bifidogenic factors. Among them lactulose is the most studied one. The nature of lactulose influence on the organism and its capability to recover and to support the growth of bifido- and lactobacteria is demonstrated by great number of research works.

The first place in a number of the main factors influential in reaction of isomerization takes the index of active acidity describing the density of protons acceptors. The biomedical research works of lactulose syrups have shown, that they are characterized by improved bifidogenous activity in comparison with conventional lactulose concentrates. The clinical research has shown the high performance of this preparation for preventive measures and treatment of a number of diseases, among them: intestine dysbacterioses, liver diseases of a different etiology, colitis and other GUT-diseases, salmonellosis, diabetes, kidney weakness, allergic responses; also used for activation of immune system etc.

Now way, most effective and scaled in industrial conditions, of the directed synthesis of lactulose is the reaction of an intramolecular lactose regrouping (L-A-transformation), catalysed by alkaline reactants. In this connection the majority of existing industrial ways of reception of lactulose preparations is based on use as raw material of lactose solutions of various high quality and concentration. Alongside with solutions of lactose for reception of lactulose preparations and various kinds of lactose-condensed dairy raw material can be used.

At the first stage of isomerization there is an intensive accumulation of lactulose at minor accumulation of by-products of reaction (the chromaticity of solutions rises). The second stage is stabilization and consequent decrease of the lactulose content; it is mostly expressed in purified lactose solutions. The increase of acidity and intensive darkening of the solution are characteristic features of this stage. Knowing chemical basics of reaction, it is possible to explain the decrease of lactulose formation. As the accumulated by-products of reaction have mostly acidic character, they start to influence the isomerization process.

The author's collective will carry out regular researches devoted to development of scientific and technical bases and technology of reception of bifidogenic concentrates of various degrees of quality. The laws of lactulose formation in various kinds of dairy raw material under action alkaline reactants are investigated The influence of technology factors (temperature, duration, pH) on an output of lactulose in modelling solutions, natural and concentrated whey is investigated. The optimum parameters of lactose isomerization process in dairy whey are established.

We carrying out now two aspects dealing with membrane fractionation for the production of lactulose:

(1) Deep fractionation of the whey, lactose and lactulose solutions
(2) Isomerization of lactose into lactulose using alkaline fraction of electrochemically activated water.

For the deep fractionation of raw materials we used baromembrane and electromembrane methods, for isomerization - electromembrane methods.

During ultrafiltration process practically all solvable calcium passes into permeate, and it is active as molasses-forming. Its salts are capable to build complexes with lactose, considerably increasing its solubility that accordingly moderates the process of crystallization in lactose production.

So we advised the membrane fractionation based on high-temperature ultrafiltration of whey. For this process cermet or polymer membranes are used. The process is held at temperature(80-90) ^{0} C. The result products are the complex of denaturized protein and solution clarified up to 80%. Thus, the penetration of membranes and the installation capacity is increased by two times compared to customary ultrafiltration.

For the increasing of clarification degree we advised the way based on permeate deacidation alongside with heating. The optimum parameters of permeates clarification are established: pH 5.8-9.8, stirring, heating up to 60-90 °C, thermostatic keeping 20 mines, centrifugal separation of the residue.

Alongside with accumulation of high-molecular substances at the surface of membrane, the reduction of its penetrating capacity is noticed due to the phenomenon of concentrating polarization. To avoid this phenomenon turbulence could be quite helpful, so we used granulated casein with the average sizes of the fragments of 0.04-0.14 mm in quantity of 0.23-1.50 kg/m^2. Thus, the degree of protein isolation increased on 3.0-3.5%, and the installation capacity increased by 2.0-2.5 times without capital costs. As a result of membranous fractionation the two products are gained: permeate with low content of protein and whey-casein paste.

The analysis of conformity of electromembranous activation process of water demonstrates that the maximum rate of pH rising can be watched only during the first 10-15 minutes (by 1-2 points pH). Then the process of proton acceptors accumulation acquires continuously decreasing character and to 60 minutes the increase of pH practically stops. We established, that for more intensive formation of hydroxide - ions it is necessary to add into the water sodium chloride or potassium chloride in quantity 0.03-0.05%.

Investigation of relaxation period, when the solution is in a metastable state and has abnormal properties, has shown, that the decrease of pH does not exceed 5% during 7 day storage (at 6 °C).

The analysis of isomerization process of lactose into lactulose at dissolution of technical lactose has shown, that the concentration of lactose in catholyte is inversely proportional to the isomerization degree of lactose into lactulose.

It is experimentally proved, that the use of higher lactose concentrations in catholyte (more than 15%) does not result in increase of isomerization degree of lactose into lactulose, as contrasted to catalytic ways. It could be explained by the shortage of alkaline ions for isomerization in catholyte and low-acid properties of lactose, causing neutralization. So further research works are devoted to the investigation of lactose solutions, subjected to electromembranous processing.

The cause of positive influencing of a mineral complex of milk on isomerization process is not only the increase of proton acceptors quantity, but also buffer properties of salts solutions permitting to support an optimum pH level at thermostatic processing of alkaline lactose solutions.

It is established, that optimum parameters of electromembranous processing are the following: concentration of lactose solution - 18-22%, voltage - 20-30 V, time about 35 minutes. The value of activation energy in this case is about 49.5 kJ/mol.

Trade marks of lactulose syrups produced using membrane fractionation are widely displayed in the Russian market. Among them "Lactulak", "Lactusan", "Alcosoft", "Lazet", “Vilakt”. Lactulose syrups gained themselves a good reputation in production of baby food and milk products - milk, yogurt, acidophilic drinks, condensed milk, ice-cream etc. The whole range of such products is produced under the trademark "Bozhja Korovka".

Lactulose is used in the products of baking and confectionery industry (bread, biscuits and sweets), in the production of soft drinks. It is included into the recipes of more than 80 types of alcoholic drinks. Functional meat products are produced with lactulose syrup: sausages, raw-smoked sausage, baby meat tinned food. So the lactulose syrups have a broad spectrum of application.

In: Biotechnology in Agriculture and the Food Industry ISBN 1-59454-119-1
Editor: G. E. Zaikov, pp. 175-180

Chapter 24

BIOTECHNOLOGICAL WAYS TO INCREASE THE FOOD VALUE PRODUCTION OF SMOKED FISH

O. Mezenova, N. Kochelaba, A. Kluchko and D. Atamanova
Kaliningrad State Technical University
236000, Kaliningrad, Sowetskij Prospekt, 1, mezenova@klgtu.ru, (0112)51-68-46

ABSTRACT

The way to increase smoked fish edible value is developed by utilization of the smoking mediums enriched natural component of plants. The edible value of such fish product is increased due to improvement of organoleptical properties, and preservation of biologically valuable substances of raw material - polyunsaturated fatty acids, essential amino acids, etc. Safety of smoked production judged on a complex of the indexes including the maintenance of biogenic amines (BA), polycyclic aromatic hydrocarbons (PAH), microbiological characteristics.

Key words: Edible value, liquid smoking mediums (LSM), natural component of plants, safety of smoked production, biogenic amines (BA), essential fatty acids.

The edible value of smoked fish production in many respects is caused, at other equal factors, from chemical composition of smoking medium.

In KSTU the way of increase of smoked fish edible value is developed by utilization of the smoking mediums enriched natural component of plants which can be practically realized by two paths.

The first path is based on introduction in liquid smoking mediums (LSM) biologically active materials of the ground plants used in the food-processing industry and pharmacology – foetus of juniper and dogrose, camomile blossom, etc., containing such active components, as essential oils, minerals and other substances useful for organism. The second direction assumes express introduction in LSM components of alga - the so-called phycocolloids having, first of all, gel-forming properties due to carragenan included in their composition.

This path allows not only enrich LSM and smoking product by useful minerals and organic matters of aqueous plants, but also improve function effect actually smoke drying.

Enriched on the first variant LSM utilized for reception of a cold smoked fish, as finished stock, and also a semiproduct for cans, prepared on type «flavored in oil». Advantage of last technology, besides improving of product quality, is absence of traditional conserving agents (benzene carboxylic acid, etc.) as components of smoking medium are provide antiseptic effect due to barrier properties.

Edible value of such fish product is increases as due to improvement of an organoleptical properties, and preservation of biologically valuable substances of raw material - polyunsaturated fatty acids, essential amino acids, etc.

So, organoleptical index of skin-off fillet of cold smoked Baltic bream, prepared with application of additives of ground plants, have been appreciated by tasters as corresponding to a specialty smoked fish: odor and taste - smoke products, with flower, coniferous and other pleasant smells; color of a surface - from be bright - golden up to dark brown, with gloss; a consistence - dense, with shades of tenderness.

One of the major index of edible value of fish products is composition of polyunsaturated fatty acids of their lipids. Polyunsaturated fatty acids are irreplaceable, causing a number of the important functions of an organism (hormone, antiseptic and so forth). In experiments it has been shown, that antioxidizing properties of smoke flavoring preparations amplify under influence of biologically active materials of the ground plants type essential oils, organic acids of plants to what the quality and quantitative fatty acid content of the lipids which have been carried out by a method of high performance chromatography (table 1) testifies. So, in control samples of smoked bream the sum of polyunsaturated fatty acids has made (in % from sum of fatty acids) 19,7 (a fillet without smoke drying) and 21,4 (a fillet with preparatVNIRO), and in experimental this index changed in a range 20,8 - 28,8.

Introduction of biologically active materials of the ground plants in smoking mediums also promotes increasing of the absolute maintenance biologically active fatty acids of ω3 families in lipids of bream.

Safety of smoked production judged on a complex of the indexes including the maintenance of biogenic amines (BA), polycyclic aromatic hydrocarbons (PAH), microbiological characteristics.

Results of chromatography of maintenance BA and corresponding loose amino acids at which decarboxylation the first collect, carried out at smoke drying on new technology, testified about biogenic safety of bream. Quantitative levels of accumulation of pentamethylenediamine, putrescin, tyramine have been established, spermidin and spermine (in limits from 9,7 up to 107,0 mg / kg of nonvolatile solid) which can be counted initial and harmless for health, thus in fish products completely there was no toxic histamine. The least harmless, that is established by calculations of express indexs of biogenic toxicity, the samples enriched by biologically active materials of the ground plants of juniper (30,4 and 141,0 %), camomiles (31,5 and 145,1 %), mints (47,9 and 241,4 %), calendula (53,6 and 217,4 %) and a linden (55,5 and 241,4 %) while bream, prepared by a traditional way of smoking, the test specimen (without smoke drying) - accordingly, 108,9 and 487,0 % it was characterized by similar indexes at a level of 68,9 and 324,0 %. Thus, follows, that biologically active materials of the ground plants are improve biogenic safety of a smoked fish on new technology.

Table 1. The maintenance of fatty acids in the lipids of cold smoked bream fillet

Fatty acids, % from $\sum$ of FA	Experimental samples (with biologically active materials of plants)									Control samples	
	Foetus of juniper	Camomile blossom	calendula blossom	Mint leaf	Lime-tree blossom	Dogrose blossom	Sea-hunter	Yarrow	min-max / average value	With smoking preparat	Without smoking
$\sum$of FA, g/100 g of dry substance	12,5	10,0	8,50	8,50	12,75	11,0	10,75	12,0	***8,50 – 12,75/ 10,63***	***11,5***	***6,00***
Saturated fatty acids (SFA)											
g/100gr of dry substance %	4,00	3,00	2,75	3,00	4,25	3,50	3,25	3,75	***2,75 - 4,25 / 3,50***	***3,80***	***2,16***
from $\sum$ of FA	31,2	31,0	31,5	33,8	34,0	31,0	31,1	32,2	***31,0 – 34,0 / 32,5***	***33,4***	***36,0***
12:0	0,1	0,2	-	-	-	-	0,1	-		-	-
14:0	3,3	3,3	3,5	4,0	3,5	2,9	3,2	3,5	***2,9 - 4,0 / 3,45***	3,8	4,3
15:0	0,8	0,9	0,8	1,0	0,8	0,6	0,7	0,9	***0,6 - 1,0 / 0,8***	0,9	1,3
16:0	20,1	19,5	20,8	21,3	22,8	21,4	21,3	21,0	***19,5 - 22,8 / 21,15***	22,2	22,6
17:0	1,2	1,3	1,3	1,5	1,3	1,0	1,0	1,2	***1,0 - 1,5 / 1,25***	1,1	1,4
18:0	4,8	5,2	4,8	5,6	5,2	4,8	4,4	5,2	***4,4 - 5,6 / 5,00***	5,1	5,9
20:0	0,4	0,4	0,3	0,4	0,4	0,3	0,3	0,4	***0,3 - 0,4 / 0,35***	0,3	0,5
22:0	0,5	0,2	-	-	-	-	0,1	-		-	-
Unsaturated fatty acids (USFA)											
g/100 g of dry substance	5,25	4,00	3,75	3,25	5,50	5,25	5,25	5,25	***3,25 - 5,50 / 4,38***	***5,20***	***2,66***
% from $\sum$ ofFA	42,8	40,2	44,3	39,3	43,0	47,2	48,1	43,3	***39,3 - 48,1 / 43,7***	***45,2***	***44,3***
14:1 ***w 6***	0,2	0,2	0,3	0,3	0,3	0,2	0,3	0,2	***0,2 - 0,3 / 0,25***	0,3	0,2
16:1 ***w 7***	10,5	8,8	11,7	9,2	10,9	11,8	14,5	10,5	***8,8 – 14,5 / 11,65***	10,4	9,9
18:1 ***w 9***	25,2	23,9	26,1	23,1	25,3	28,9	26,9	25,5	***23,1 - 28,9 / 26,0***	28,7	27,2
18:1 ***w 9***	6,0	6,3	5,4	5,5	5,5	5,6	5,6	6,3	***5,4 - 6,3 / 5,85***	4,9	5,7
20:1 ***w 9***	0,9	1,0	0,8	1,2	1,0	0,7	0,8	0,8	***0,8 - 1,2 / 1,00***	0,9	1,3

Table 1. Continued

Fatty acids, % from ∑ of FA	Experimental samples (with biologically active materials of plants)									Control samples	
	Foetus of juniper	**Camomile blossom**	**calendula blossom**	**Mint leaf**	**Lime-tree blossom**	**Dogrose blossom**	**Sea-hunter**	**Yarrow**	**min-max / average value**	**With smoking preparat**	**Without smoking**
Polyunsaturated fatty acids (PUSFA)											
g/100 g of dry substance %	3,25	3,00	2,00	2,25	3,00	2,50	2,25	3,00	***2,25 - 3,25 / 2,75***	***2,50***	***1,18***
from ∑ ofFA	26,0	28,8	24,2	26,9	23,0	21,8	20,8	24,5	***20,8 - 28,8 / 24,8***	***21,4***	***19,7***
18:2 ***w 6***	4,0	5,3	3,9	4,4	3,7	3,3	2,9	3,7	***2,9 - 5,3 / 4,1***	3,2	3,1
18:2 ***w 6***	0,4	0,4	0,6	0,4	-	0,4	0,4	0,4	***0,4 - 0,6 / 0,5***	0,6	0,2
18:3 ***w 3***	2,8	3,4	2,3	2,9	2,8	2,6	1,8	2,7	***1,8 - 3,4 / 2,6***	2,3	2,1
18:4 ***w 3***	0,3	0,4	0,4	0,3	0,3	0,5	0,2	0,5	***0,2 - 0,5 / 0,25***	0,2	0,2
20:2 ***w 6***	1,6	1,6	1,2	1,9	1,5	1,1	1,2	1,7	***1,2 - 1,9 / 1,55***	1,4	1,6
20:4 ***w 6***	3,0	3,2	2,8	3,4	3,0	2,1	2,5	3,0	***2,1 - 3,4 / 2,75***	2,5	2,4
20:5 ***w3***	6,6	7,4	7,2	6,6	5,7	5,7	6,2	6,6	***5,7 - 7,4 / 6,55***	5,8	5,3
22:5 ***w3***	1,7	1,7	1,6	2,1	1,5	1,3	1,5	1,6	***1,3 – 2,1 / 1,7***	1,5	1,7
22:6 ***w 3***	5,6	5,4	4,2	4,9	4,5	4,8	4,1	4,3	***4,1 - 5,6 / 4,85***	3,9	3,1
∑ unsaturated fatty acids											
g/100 g of dry substance	8,50	7,00	5,75	5,50	8,50	7,75	7,50	8,25	***5,75 - 8,50 / 7,13***	***7,70***	***3,84***
% from ∑ ofFA	68,8	69,0	68,5	66,2	66,0	69,0	68,9	67,8	***66,0 – 69,0 / 68,5***	***66,6***	***64,0***
∑ w3 FA., % from ∑ ofFA	16,7	18,3	15,7	16,8	14,8	14,9	13,8	15,7	***13,8 – 18,3 / 16,05***	***13,7***	***12,4***
∑ w6 FA., % from ∑ ofFA	9,2	10,7	8,8	10,4	8,5	7,1	7,3	9,0	***7,1 – 10,7 / 8,9***	***8,0***	***7,5***
w3 FA / w6 FA	1,82	1,71	1,78	1,62	1,74	2,10	1,89	1,74	***1,62 - 2,10 / 1,86***	***1,71***	***1,65***
PUSFA/ USFA	0,81	1,00	0,73	0,75	0,71	0,71	0,69	0,80	***0,69 – 1,00 / 0,85***	***0,66***	***0,55***

Ecological safety of smoke products judged under the maintenance in them PAH, getting in a fish extremely from smoking mediums. The data obtained by a method of high performance solution chromatography, are given in table 2.

Table 2. The maintenance of PAH in a cold smoked bream fillet, mkg / kg

Compounds	Degree of carcinogenic activity	Fillet of bream prepared with application of LSM with biologically active materials of plants		
		camomile	mint	linden
Fenantren		1,9	0	1,4
Piren		0,7	0,5	0,4
Хризен	+	1,2	0	0
Fluoranten		0,7	0,8	0
Benz(a)antracen	+	2,1	0	0
Benz(b)fluoranten	++	0	0	0
Benz(k)fluoranten		0	0	0
Perilen		0	0	0
Benz(a)piren	+++	0	0	0
Benz(e)piren		0	0	0
Benz(q,h,i)perilen		0	0	0
Bibenz (a,c)antracen	+	0	0	0
Bibenz (a,h)antracen	+++	0	0	0
Bibenz (a,i)piren	+++	0	0	0
Bibenz (f,e)piren		0	0	0
Bibenz (a,h)piren		0	0	0
Koronen		0	0	0
Fluoren		0	0	0
Antracen		3,4	0	0
Indeno(1,2,3-c,d)piren	+	0	0	0
Σ		10,0	1,3	1,4

* +++; ++; + - accordingly high, average and low cancerogenic activity

In microbiological studies the increasing of antiseptic safety of the smoked production prepared with application new LSM is confirmed, that also it is possible to explain composition introduction natural biologically active materials of the plants (organic acids, ethers, terpenic hydrocarbons, etc.). So, by the end of a normative shelf-life of bream (45 day) in the experimental sample processed with LSM enriched juniper additives and control, prepared on the base of smoke preparat VNIRO and without smoking ingredients made values of index of bacteria number accordingly 50, 300 and 3000 cell/g, that did not exceed a specified level (104 cell/g). Pathogenic forms (coli-form bacteria) in the volumes regulated by the normative documentation, are not found out.

LSM, prepared by enrichment of smoke preparat VNIRO by biologically active materials of alga of Baltic sea of sort Rhodohpyta, have been investigated in technology of cool smokeless smoking fish on immersion principle. Introduction of phycocolloids of carragenan type, containing besides carragenan alginic, staining and number of accompanying extractives of alga, has allowed to receive viscous smoking gel which formed characteristic smoking "film" on a surface of a fish after drying. This «film», besides the basic function purpose, has increased appearance of production and preserving effect.

Comparative researches of quality of cold smoked fishes with carragenan which have been carried out concerning prepared «on starch» under technical specifications 15-1207-96, has shown the increasing of physical and acid stability of experimental "film", amplification natural brown color, and also propagation of antiseptic effect of smoking as a whole. At introduction of phycocolloids it is possible to explain amplification of function properties of smoking gels by the nature of carragenan which contain number of sulphur group. It is allows them to develop not only gel-forming, but also surface-active, antiseptic, antifungal and other important properties. Researches of structural-mechanical properties of such gel, carried out on measuring instrument of viscosity PAAR PHYSICA have shown, that maximal value of viscosity makes 0,108 Pa*c.

The antiseptic properties of new smoking medium investigated during preparation and storage on a fillet of cold smoked mackerel concerning control, have authentically shown the increasing of microbiological safety of enriched production. So, right after smoking the level of index bacteria number in production dropped with $9{,}65*10^5$ cell/g (in raw material) up to $1{,}4 * 10^4$ and $2{,}8*10^2$ cell/g at the samples prepared with utilization accordingly of starch and phycocolloids. During storage microorganisms continued to die off and value of bacteria number after 7 weeks dropped accordingly up to 89 and 21 cell/g. Thus the sanitarian - indicative microflora was absent during all period of researches.

The carried out research testifies that the way of introduction in standard smoking mediums composition of biologically active materials of plants and alga describes by the high maintenance of biologically active substances, allows to obtain the liquid smoking mediums distinguished by improving of function properties, preserving effect, safety, and their application in technologies of fish product promotes increasing of its edible value and quality as a whole.

References

[1] L.Toth: *Chemie der Räucherung*. Verlag Chemie, Bonn, 1983. 331 p. (in Germany)

[2] O. Mezenova, I. Kim, S. Bredihin: *Production of smoked products*. Kolos, Moscow, 2001. 208 p.

In: Biotechnology in Agriculture and the Food Industry
Editor: G. E. Zaikov, pp. 181-184
ISBN 1-59454-119-1

Chapter 25

CRITERIA FOR ASSESSING QUALITY AND PRESERVATION OF CONFECTIONERY

Alla Vladimirovna Ryzhakova
Research Institute of the Russian Plekhanov Academy,
Assistant Professor of Commodity Expertise and Trade Department,
Moscow, Russia, SRD@rea.ru

ABSTRACT

Results of researching Russian confectionery aimed at enhancement of quality and durability extension are presented.

Key words: quality of confectionery, storage processes, durability.

INTRODUCTION

Confectionery is a group of food products of rather wide range, which have high nutrition value, excellent taste and pleasant appearance. It is notable for specific taste and odour, composition, appealing and decorative appearance. In 2002 the output of confectionery in Russia made up 1,800,000 tons, annual consumption per head is 12-14 kg.

Confectionery is not basic food, but taking into account its significance in the life of people it plays an important role in sustaining power balance. Today confectionery branded "Made in Russia" is popular on the confectionery market of Russia.

Sweets with fondant body takes a special place among confectionery. It should be noted that these sweets are in great demand with low- and middle- income customers.

EXPERIMENTAL

Current ideas concerning enhancement of sweets' quality covers the issue of products' getting stale during storage, this process is accompanied by worsening of the structure, taste and appearance. White spots emerge on the product's surface and on its break, they are clusters of microscopic crystals of saccharose.

We should point out that the issue of increasing durability of fondant sweets during storage was discussed in many papers but the mechanism of 'getting stale' has not been understood yet.

Speed of products' getting stale depends mainly on storage conditions and presence of substances which can stop evaporation and saccharose crystallization, which includes in particular, inert matters.

Fondant sweets made with the use of highly-maltose treacle of three types (samples 1, 2 and 3) and with bio-invert (sample 2) were researched. The check sample was a sample made according to the traditional recipe.

To study the processes which take place during six months' storage of fondant sweets organoleptic figures were estimated and physical and chemical parameters were defined: content of reducing matters, humidity, ash content of fondant bodies, content of icing and activity of water a_w.

Organoleptic figures (taste and odour, condition of the surface, shape, structure) were estimated on the basis of 5-grade scale with regard for weighting factors and value of separate parameters.

The following weighting factors were fixed:

taste and odour	8
shape	1
appearance	4
structure	7

Weighting factors of quality indicators were taken into account at the stage of test sheets' processing and calculating the generalized quality indicator.

The results of complex indicators of quality are shown in Graph 1. As we can see in the Graph fresh samples of sweets have the same level of quality, but when storing periods fixed by regulatory documents expire (4 months) we can see the obvious advantage of sweets made with the use of highly-maltose treacle and bio-invert, which was confirmed by subsequent storage up to 6 months.

It was defined that samples made with the use of highly-maltose treacle and bio-invert had much better commodity expertise features as compared with the check sample, while the use of food additives (sample 2) allows to stabilize quality figures of sweets.

Results of test assessment comply with the data of physical and chemical analysis, water activity is much lower in test samples of sweets than in check ones. It is well-known that increasing activity of water in the product by 0.01 halves storing periods (1).

Thus, it was found out that the use of a new type of treacle for production of fondant sweets and the invert matter on the basis of invertase can both increase the storing period of products and predict storage by the indicator – water activity.

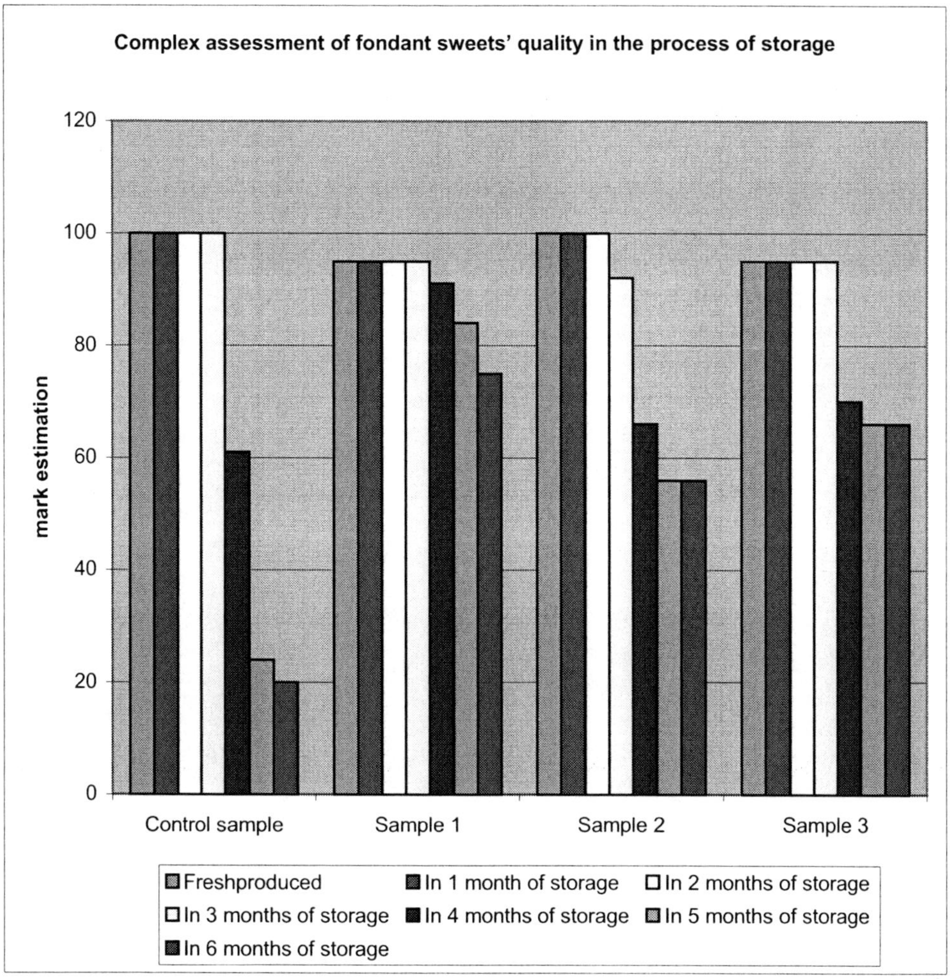

The product range of praline sweets is wide enough, but they have a short storing period (according to Russian standards). Therefore, today specialists' efforts are focussed on enhancement of products' quality and their preservation. One of the most promising ways of solving the problem is the use of natural antioxidant, Tokomix.

Praline sweets without icing made by the Russian factory "Rot Front" (confectionery sticks) became the object of the research. The sweets are cylindrical, they are made of finely crushed mixture of fried kernels of peanuts containing oil, sugar and grease.

Four samples of sweets (three batches) were examined, they were given the following names:

Sweets made by "Rot Front" (check sample)	sample made according to traditional recipe;
Sample N 1	sweets containing 0.2% of antioxidant Tokomix;
Sample N 2	sweets containing 0.4% of antioxidant Tokomix;
Sample N 3	sweets containing 0.6% of antioxidant Tokomix.

The quality of products was estimated on the basis of organoleptic figures, moisture content, peroxide and acid number, carbonyl combination and microbiological indicators during 6 months.

Results

Test assessment showed quality of "Rot Front" sweets did not change considerably. All samples after 3 months' storage had good taste, soft composition and fine-crystalline homogeneous structure. After 4 months' storage changes took place in check samples and samples with 0.2% of Tokomix.

Samples with 0.4% and 0.6% of Tokomix were found excellent. After 6 months' storage the sample with 0.6% of Tokomix was announced the best. Moisture loss for the period of research (4 months) in sweets made by "Rot Front" reached: in the check sample – 0.91% and in samples 1,2 and 3 – 0.72; 0.43 and 0.26, correspondingly. Thus we can made a conclusion that presence of antioxidant Tokomix in sweets' recipe improves their water-retaining capacity.

Quality stability in praline sweets without icing in the process of storage depends mainly on resistance of the praline substance to hydraulic oxidazing processes.

Slight increase in acid and peroxide number after 4 months' storage testifies to the high quality of imported vegetable oil. Addition of antioxidant allows to reduce the accumulation of carboxyl combinations during storage. Positive effect of antioxidant is obvious.

It should be noted that the level of microbiological indicators can also guarantee customers' safety.

Conclusion

Summing it up we should underline that the quality of Russian confectionery is one of the most important competitive advantages on the world market. The leading Russian producers have retained and now support scientific and technological positions in respect of these parameters. At the same time we can guarantee quality only in case constant attention and scientific support are provided in the field of shaping and preserving quality of products.

References

[1] Reichart O., Lehoczki-Tornai J. Effect of water activity on the kinetics of destruction of Esclerichia coli by chemicals. *Acta alim*, 21, N2, 1992.

In: Biotechnology in Agriculture and the Food Industry ISBN 1-59454-119-1
Editor: G. E. Zaikov, pp. 185-198

Chapter 26

IMMUNOMODULATING AND ANTITUMOR ACTION OF EXTRACTS OF SEVERAL MUSHROOMS

Mark Shamtsyan[1], Valentina Konusova[2], Natalya Pigareva[2], Alexander Goloshchev[1], Yulia Maksimova[1], Andrey Panchenko[3], Andrey Simbirtsev[2], Nikolay Petrishchev[3] and Nina Denisova[4]

[1]Department of Technology of Microbiological Synthesis, St.Petersburg State Institute of Technology (Technical University), 26 Moscovsky Pr., 198013 Saint Petersburg, Russia, e-mail: shamtsyan@yahoo.com;

[2]Laboratory of Immunopharmacology, State Research Institute of Highly Pure Biopreparations, 7 Pudozhskaya Str., 197110 Saint Petersburg, Russia;

[3]Department of Pathological Physiology, Acad.I.P.Pavlov St.Petersburg Medicinal University, 6/8 L.Tolstoy Str.,197022 Saint Petersburg, Russia;

[4]Laboratory for Biochemistry of Mushrooms, V.L.Komarov Botanical Institute, Russian Academy of Sciences, 2 Prof.Popov Str., 197376 Saint Petersburg, Russia

ABSTRACT

Aqueous extracts from fruit bodies and mycelia of various Higher Basidiomycetes were studied for reliable biological effects. It was determined that investigated preparations are increasing production of reactive oxygen forms by neutrophils of peripheral human blood, possessing marked mitogenic activity in a wide range of concentrations and stimulating production of inflamatory cytokines interleukine 1-β and interleukine – 8 by peripheral blood cells. Oyster mushroom extracts were also studied for antitumor activity. It appears, that their oral administration is prolonging the survival rate of mice with transplanted melanoma B16 and decreasing both average tumor sizes and manifestation of tumor intoxication.

Key Words: Higher Basidiomycetes, immunomodulation, anti tumor effect

INTRODUCTION

It is well known that Higher Basidiomycetes have a wide spectrum of therapeutic and prophylactic activities. About semi-centennial way was covered by modern science from the first studies of antitumour activity of extracts from carpophores of various species of macromycetes [1-3] till appearance of "basidial" remedies in the world pharmaceutic market [4, 5].

Last decades the strategy of investigations of the antitumour and immunomodulating features in high basidiomycetes logically disembogued in the general course of complex research of therapeutic, prophylactic, nutritious and other benefits of macromycetes. In the first place it had respect the most popular species in mushroom industry from genus *Agaricus, Pleurotus, Lentinus, Ganoderma, Grifola, Flammulina, Volvariella, Auricularia, Tremella* [6]. It is necessary to mention that information about popular market species of mushrooms are uneven. Thus, the species *Lentinus edodes* or *Ganoderma lucidum* have a long history of careful research while intensive investigation of other taxons was started only at the second half of passed centenary. Species of genus *Pleurotus* can be mentioned as example of the latter. In the contrary, several other species, like *Innonutus radiatus, Tricholoma partentosum* or *Bjerkandera adusta,* treated among the others in our studies were never tested before for immunomodulating activity.

We would like to represent the experimental investigation of immunomodulating activity of several mushrooms: *Armillaria mellea* (Vahl: Fr.) P. Kumm., *Bjerkandera adusta* (Willd.: Fr.) P. Karst., *Flammulina velutipes* (Curt.: Fr.) Singer, *Fomes fomentarius* (L.: Fr.) Fr., *Fomitopsis pinicola* (Sw.: Fr.) P. Karst., *Ganoderma lipsiense* (Batsch) G.F. Atk. (= G. applanatum (Wallr.) Pat.), *Piptoporus betulinus* (Bull.: Fr.) P. Karst., *Pleurotus cornucopiae* (Pau. ex Pers.) Rollan (= P. citrinopileatus Singer), *P. ostreatus* (Jacq.: Fr.) P. Kumm., *Trametes suaveolens* (Fr.) Fr., *Auricularia auricula-judae* (Bull. ex St. Amans) Wettst., *Inonotus obliquus* (Pers.: Fr.) Piliÿt, *Inonotus radiatus* (Sowerby: Fr.) P. Karst., *Tremella foliacea* Pers.: Fr., *Tricholoma portentosum* (Fr.) Quil., *Grifola frondosa* (Dicks.: Fr.) Gray, *Lentinus edodes* (Berk.) Singer, as well as anti tumor activity of *Pleurotus ostreatus.*

MATERIALS AND METHODS

Materials

Objects of our study were aqueous extracts of fruiting bodies and mycelium of mushrooms of species *Armillaria mellea*, *Bjerkandera adusta*, *Flammulina velutipes*, *Fomes fomentarius*, *Fomitopsis pinicola*, *Ganoderma lipsiense*, *Piptoporus betulinus*, *Pleurotus cornucopiae*, *Pleurotus ostreatus*, *Trametes suaveolens*, only fruit bodies extracts of species *Auricularia auricula-judae.*, *Inonotus obliquus*, *Inonutus radiatus*, *Tremella foliacea*, *Tricholoma portentosum*, and also extracts from cultural mycelium of species *Grifola frondosa* and *Lentinus edodes*.

Mycelium was gained by submerged cultivation of the strains on the wort (4°B) at 28°C, dryed at 60°C, crumbled up and treated with 85% ethanol at 80°C during 8 h. Sediment was twice extracted by boiling water during 3 h. The obtained filtrate was boiled out under vacuum, polysaccharide fractions were precipitated by 5 volumes of ethanol, dialyzed against distilled water and freeze dried. The obtained preparations were tested on immunomodulating and antitumoural activities.

Reagents

For immunological investigations fresh donor blood, phycoll-pack (d 1,078 kg m^{-3}), Henks salt solution ("Biolot"), luminol, phorbolmiristate acetate (PMA), agarose EEO (type 11), formyl-peptide fMLP 1640, cow fetal serum, prodigiozan, concanavalin A (all "Sigma") were used.

Blood Cells Isolation

Obtaining mononuclear and granulocyte cells from donor blood stabilized with heparin solution (20 ME per 1 ml of blood) was made with standard technique described by Boyum– in phycall-pack gradient [7]. Blood plasma was layered to phycall-pack gradient and centrifuged during 40 min on 400 g. For obtaining mononuclears the cells were collected from the interphase between plasma and phycall-pack. Neutrophilic granulocytes were obtained from the precipitate after isolation of mononuclear cells by a hypotonic shock with distilled water during 30 s.

Study of the Proinflamatory Cytokin-Inducing Activity of the Preparations

Mononuclear cell suspension (concentration 5 10^6 ml^{-1}) resuspended in RPMI 1640 medium and included 10% fetal serum was placed in the 96-holes plainbottom planchette ("Costar") by 0.1 ml in each hole. Simultaneously, the investigating preparations in appropriate solutions were added in the same holes. The Russian preparation of lipopolysaccharide nature "prodigiosan", a strong inductor of proinflammatory cytokines was used as the positive control. The planchettes were placed in CO_2-incubator during 24 h, after that the amount of cytokines in the cultural medium was detected by immunoenzymatic method [8]. Test-systems made by the company "Cytokine" (Saint Petersburg, Russia) were used.

Study of the Interleukine-2 Inducing Activity of the Preparations

To study the influence of extracts on the production of interleukine-2 (IL2) the mice splen cells ($5x10^6$ in 1 ml) were cultivated in RPMI-1640 medium (Sigma) with the addition of 5% FCS, 2MM L- glutamine (Sigma), 80 μ/ml of gentamicine and 1 μ/ml KonA (Serva) at the presence of various concentrations of studied preparations (concentration range from 100 to 1,56 μ/ml). Cells were cultivated for 24 hours at 37° C in the atmosphere containing 5% CO_2. At the end of the cultivation from the each hole of the plate 150 μ of supernatant was taken and kept in refrigerator at -70° C until further treatment.

The content of IL2 in the supernatants was determined in biological test using IL2 dependent line of CTLL2.

Study of the Preparations Influence on the Production of Oxygen Active Forms by Phagocyting Cells

The formation of oxygen active forms by whole blood cells was detected with luminoldependent chemiluminescence method (LDCL) [9]. The determination of LDCL was carried out in the 96-holes white untransparent planchettes "Costar" with the device Victor-2 (Finland), connected with PC and provided sequent registration of chemiluminescence. The reaction result was termed as light sum (sum of impulse accumulated during the certain time period).

The base luminol solution (10^{-2} M) was prepared in DMSO and stored at 4°C. Working solution (10^{-4} M) was preparing just before using through diluting the base solution with Henks solution (pH 7.2).

The experiment was carried out follow scheme: 120 mkl of the Henks solution, 20 mkl of the fresh donor blood, 40 mkl of the working luminol solution and 20 mkl of the investigating preparation in the appropriate dilutions were placed in the each hole of the 96-holes planchettes. Every investigation was prosecuted in the three parallels. The solution of phorbol-miristate acetate (PMA) in the end concentration 10 ng ml^{-1} was used as the standard positive control. The reaction was fixed during 1 h.

The Preparations Influence on the Cell Proliferation

The influence of mushroom extracts on the proliferation of thymus and spleen cells was studied in the reaction of blasttransformation of lymphocytes of CBA strain mice (farm Rappolovo, Saint Petersburg, Russia). The thymus cells (final concentration $5x10^{6}$) and spleen cells (final concentration $2,5x10^{6}$) were cultivated in the culture medium RPMI-1640 (Sigma) with the addition of 2mM of L- glutamine (Sigma), 80 μ/ml of gentamicine and 5% FCS, inactivated by heating at 56°C for 40 minute on the background of various concentrations of studied preparations (concentration range from 100 to 1,56 μ/ml). Konkanavaline A (Serva) (KonA) in the final concentration of 0,5 μ/ml or lipopolysaccharide (Sigma) in concentration of 10 μ/ml were used to stimulate proliferation of thymus and spleen cells. To performe the reaction of blasttransformation the 96-hole planchettes for cultivation (Costar) were used. Cell culture was incubated at 37° C in the atmosphere containing 5% CO_2 for 72 hours. 16 hours before the end of cultivation period the labeled tritium ^{3}H thymidine (Isotope) at the final concentration of 200 μBK per ml was added into the all holes with cultural medium. After cultivation the amount of accumulated in the cells ^{3}H thymidine was determined using liquid scintillation counter (Rackbeta 1217). Results were expressed in impulses per minute.

Results and Discussion

Primary evaluations of biological activity for aqueous extracts of studied mushrooms were done by test-analysis of the reactive oxygen species generated by cells of a peripheral human blood by a method of luminoldepended chemiluminescence. The test was selected on the basis of the supposition about stimulating effect of extracts on the system of phagocytes, since the main components of extracts were determined to be polysaccharides, and it is known, that the compounds of polysaccharide nature boost the phagocytosis. When activating the phagocytes of the blood the cascade of chain reactions is started accompanied with the increased oxygen consumption from the environment. Thus the cells pass on hexosomonophosphate path of respiration. The final products of the reaction contained the reactive oxygen species, which being combined with luminol caused a luminescence registered at a wave-length 425 nm.

The extracts from fruit bodies of all mentioned species were studied in concentrations 0.001-1.0 mg ml^{-1} (table 1). On a background of PMA as a positive control many of the experienced samples of mushrooms increased production of reactive oxygen species by neutrophils of a peripheral human blood. The maximal effect of all experienced extracts observed in a dose at 0.1 - 1.0 mg ml^{-1}. Comparison of the influence of fruit bodies extracts with the influence of mycelial extracts of same species shows, that the stimulating effect of fruit bodies extracts was slightly higher, then those observed for mycelia (table 2). In both cases the maximal effects were observed for extracts of mushrooms of P. ostreatus and P. cornucopiae

Table 1. The effect of extracts from mushrooms fruit bodies on reactive oxygen species production by human blood neutrophils

Investigating preperations	Preparation concentrations, mg / ml			
	1	0,1	0,01	0,001
	Light sum, imp s^{-1}			
Extract from Armillaria mellea	77 ± 5	71± 3	53± 3	39± 6
Extract from Bjerkandera adusta	267± 7	355± 4	211± 2	88± 2
Extract from Flammulina velutipes	83± 4	72± 6	59± 3	40± 2
Extract from Fomes fomentarius	63± 2	55± 1	47± 3	41± 3
Extract from Fomitopsis pinicola	59± 5	45± 2	40± 1	36± 2
Extract from Ganoderma lipsiense	90± 8	115± 9	73± 8	48± 8
Extract from Piptoporus betulinus,	56± 4	43± 3	38± 7	40± 2
Extract from Trametes suaveolens	98± 3	112± 3	75± 5	53± 3
Extract from Pleurotus ostreatus	380± 12	392± 7	128± 8	87± 5
Extract from Pleurotus cornucopiae	371± 8	383± 4	188± 7	111± 6
Extract from Auricularia auricula-judae	350± 6	370± 8	250± 5	120± 7
Extract from Inonotus obliquus	52± 2	45± 4	39± 3	35± 2
Extract from Inonotus radiatus	125± 3	175± 6	108± 7	60± 2
Extract from Tremella foliacea	117± 5	152± 9	121± 3	83± 1
Extract from Tricholoma portentosum	236± 11	139± 2	115± 5	65± 3
PMA (positive control)			179± 8	
background (whole blood)		35± 3		

The mitogenic activity of fruit bodies extracts for mushrooms which possessed higher activity in generation of the reactive oxygen was studied. The research was carried out in blast-transformation reaction (BTR) on spleen and thymus cells of CBA mice.

Table 2. The effect of extracts from mushroom mycelium on the production of reactive oxygen species by human blood neutrophils

Investigating preperations	Concentrations of preparations, mg / ml	
	1	0,1
	Light sum, imp s^{-1}	
Extract from Armillaria mellea	65± 8	58± 8
Extract from Bjerkandera adusta	243± 10	325± 12
Extract from Flammulina velutipes	71± 6	63± 7
Extract from Fomes fomentarius	52± 3	45± 3
Extract from Fomitopsis pinicola	48± 2	43± 2
Extract from Ganoderma lipsiense	78± 4	92± 5
Extract from Piptoporus betulinus	45± 2	42± 3
Extract from Trametes suaveolens	87± 7	102± 4
Extract from Grifola frondosa	284± 9	291± 4
Extract from Lentinus edodes	338± 9	340± 5
Extract from Pleurotus ostreatus	353± 8	369± 9
Extract from Pleurotus cornucopiae	335± 5	343± 7
PMA, 0,1 mg ml-1 (positive control)	185± 8	
background (whole blood)	38 ± 2	

It is clear from the table 3 that the extracts of all tested mushrooms have marked mitogenic activity in a wide range of concentrations. In this tests the maximal activity levels were detected for the extracts of *B. adusta*. Both species of *Pleurotus* mushrooms were also active in a wide range of concentrations.

Activation of thymus cells by tested preparations was negligible.

Besides the mitogenic activity of fruit bodies, the same activity of mycelia of B. adusta, P. ostreatus and P. cornucopiae also was studied.

All the preparations were possessing expressed influence on the spontaneous proliferation of spleen cells (Fig. 1), however the level of this action significantly differs: thus, preparation from mycelium of B. adusta was causing reliable increase of spleen cell proliferation only in the dozes of 100 and 50 μ/ml, preparation from mycelium of P. ostreatus – in the dozes from 100 to 12,5 μ/ml, while preparation from P. cornucopiae was significantly increasing the proliferation of spleen cells in all of the studied dozes. The efficiency of mitogenic action of the last preparation is comparable with the influence of well known mitogen – lipopolysaccharide. Various concentrations of tested preparations being simultaneously added to the spleen cell culture with suboptimal dozes of lipopolysaccharide (0,5 μ/ml) do not possessне neither stimulating, nor inhibiting action. However simultaneous addition of the mycelial preparations to the spleen cell culture with suboptimal dozes of KonA (0,5 μ/ml) demonstrates an inhibiting action of the extract from B. adusta in the dozes of 100 and 50 μ/ml (Fig. 2).

Table 3. Influence of mushroom extracts on blast-transformation reaction of lymphocytes from CBA mice spleen

Extracts from fruit bodies of mushroom species	Preparation concentration, mg / ml						
	0.5	0.25	0.125	0.0625	0.03125	0.0156	0.0078
	Level of blust-transformation reaction, thousands imp min^{-1}						
Bjerkandera adusta	76.1± 0.4	80.6 ± 0.7	74.4 ± 0.3	55.1 ± 0.4	48.3 ± 0.8	42.6 ± 1.1	38.1 ± 0.7
Inonotus radiatus	22.7 ± 0.5	23.5 ± 0.2	23.8 ± 0.5	19.3 ± 0.5	12.9 ± 0.3	10.7 ± 0.6	8.7 ± 0.3
Ganoderma lipsiense	15.9 ± 0.2	15.1 ± 0.3	17.4 ± 0.3	11.9 ± 0.3	9.5 ± 0.5	8.9 ± 0.4	6.3 ± 0.2
Tricholoma portentosum	27.9 ± 0.7	29.5 ± 0.2	26.9 ± 0.9	22.1 ± 0.3	16.0 ± 0.7	12.0 ± 0.5	11.0 ± 0.3
Pleurotus ostreatus	53.0 ± 0.2	53.9 ± 0.2	53.3 ± 0.5	49.5 ± 0.4	44.8 ± 1.0	45.6 ± 0.8	31.5 ± 0.3
Pleurotus cornucopiae	22.5 ± 0.2	39.3 ± 0.3	45.2 ± 0.5	45.8 ± 0.5	49.0 ± 0.7	40.0 ± 0.3	29.6 ± 0.8

Level of spontaneous BTR - 8.3 thousands imp min^{-1}
Level of BTR as response on ConA (0.1 mg ml^{-1}) – 32.3 thousands imp min^{-1}

Table 4. Influence of an extracts of mushroom fruit bodies on production of inflamatory cytokines by human blood cells

Concentration of preparations, mg/ml	Extracts from mycelia						Extracts from fruit bodies			
	B.adusta		P. ostreatus		P. cornucopiae		I. radiatus		T. portentosum	
	Level of cytokine production, ng ml^{-1}									
	IL 1-β	IL-8	IL 1-β	IL-8	IL 1-β	IL-8	IL 1-β	IL-8	IL 1-β	IL-8
5.0	2155	6753	2216	3178	1420	4773	2455	>10000	1882	7874
0.5	1438	3911	1879	2950	1844	6797	1820	>10000	2231	>10000
0.05	916	2773	1287	2877	1514	3987	968	9006	1748	3736
0.005	709	2230	1015	2383	1021	2440	813	4196	1275	2250
0.0005	671	2170	815	2161	872	2180	710	2877	898	2160

Level of spontaneous production of IL 1- β: 670 ng ml^{-1}, IL-8: 2169 ng ml^{-1}
Level of production induced by LPS of IL 1- β: 1653 ng ml^{-1}, IL-8: 6668 ng ml^{-1}

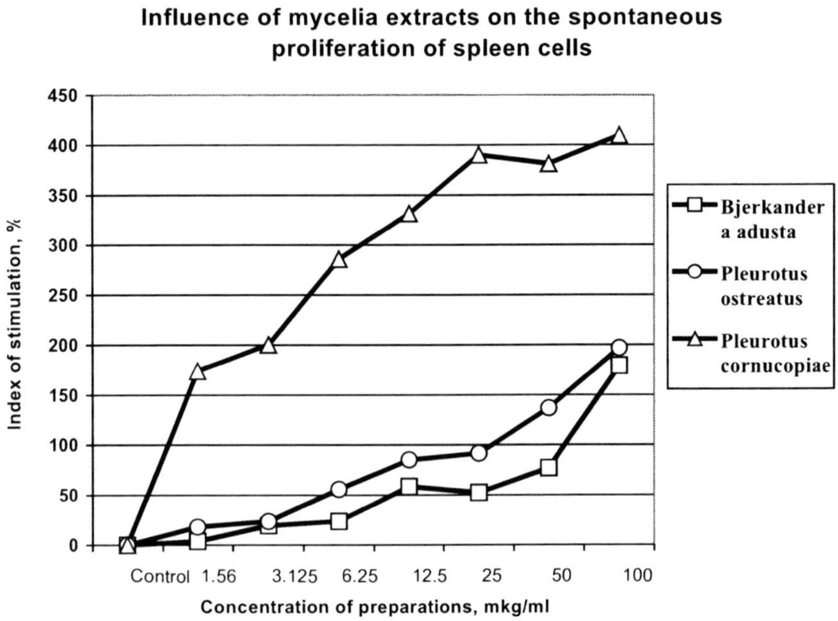

Figure 1.

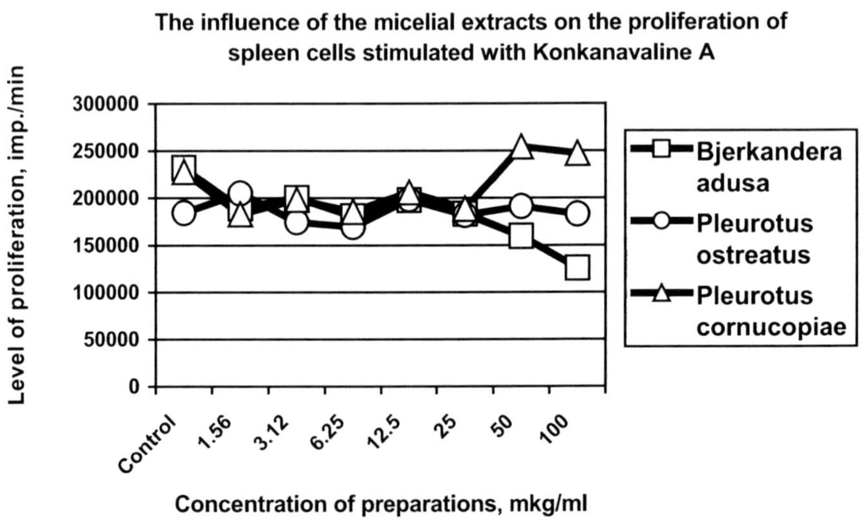

Figure 2.

Preparations do not influence spontaneous proliferation of thymus cells, except the extract from P. cornucopiae, which was reliably increasing their proliferation in the dozes from 100 to 12,5 μ/ml (Fig.3). However, when this preparation was used with thymus cell culture on the background of their stimulation with KonA, no influence on the proliferation was detected. (Fig.4). Extract from P. ostreatus micelium was reliably inhibiting KonA stimulated proliferation of thymus cells in the dozes from 100 to 25 μ/ml. Preparation from B. adusta was inhibiting the proliferation of thymus cells in the wide range of dozes: from: 100 μ/ml to 6,25 μ/ml.

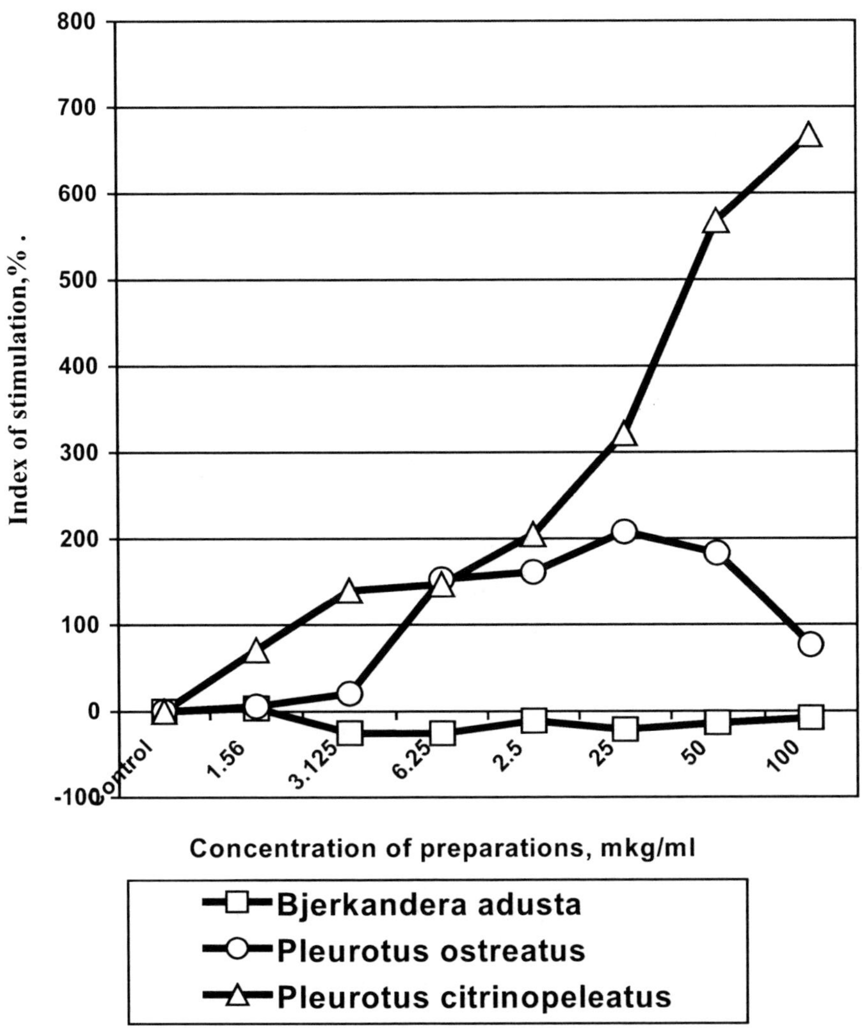

Figure 3. The influence of micelia extracts on the spontaneous proliferation of spleen cells

It was studied the ability of extracts from fruit bodies and mycelium of several mushrooms to participate in synthesis by human blood cells of immunoregulating proteins – cytokines, particularly of interleukine-1β and interleukine- 8, being a key mediator of inflammatory and immune response of an organism. The ability of a mushroom extracts to influence lipopolysaccharide induced cytokine production was studied. The data obtained are shown in table 4.

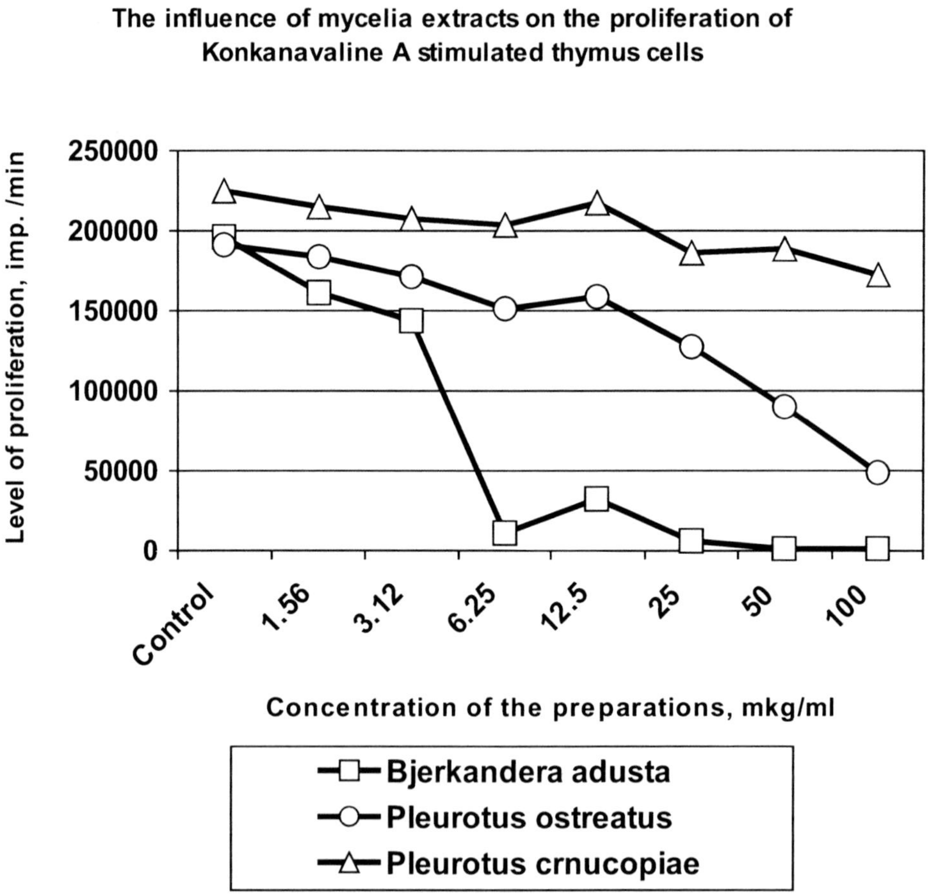

Figure 4.

It can be seen from the table 4 that preparations display a costimulating effect on both cytokines production by peripheral blood cells. More effective were preparations *from I. radiatus* and *T. portentosum,* which represents fruit body extracts, in contrary to other preparations, which were extracts from submerged mycelia.

On the Fig. 5 and 6 the results, obtained in the study of production of IL-2 by the spleen cells of mice are represented. No spontaneous activation of IL-2 production by studied preparations was detected (Fig. 5). Most of the preparations have no influence also on the production of IL-2 by Konkanavaline A stimulated spleen cells, while the extract from mycelium of B. adusta significantly inhibits Il-2 production in the dozes from 100 to 25 μ/ml (Fig 6). It is not possible to explain now this phenomena, because in our studies we deal with not enough purified preparations.

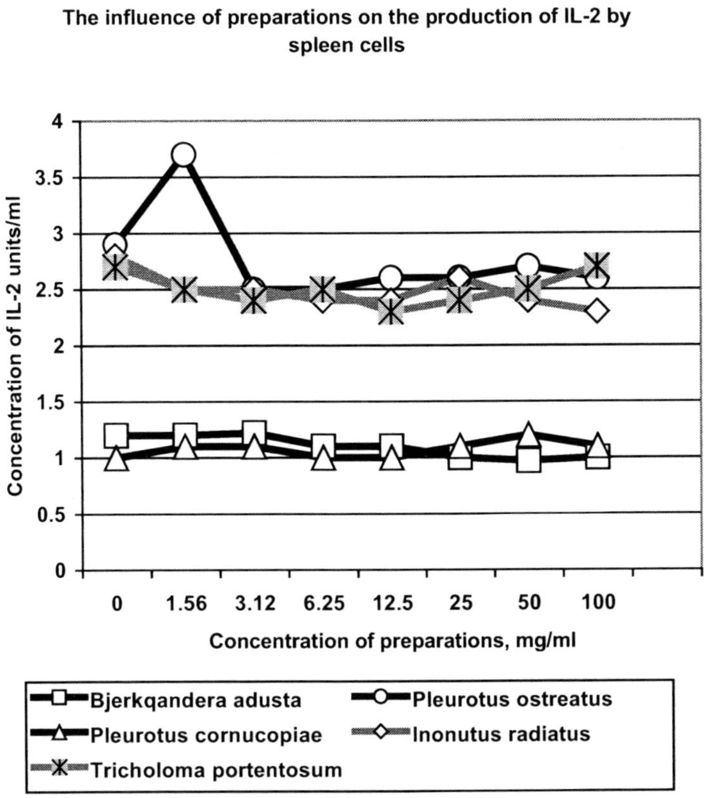

Figure 5.

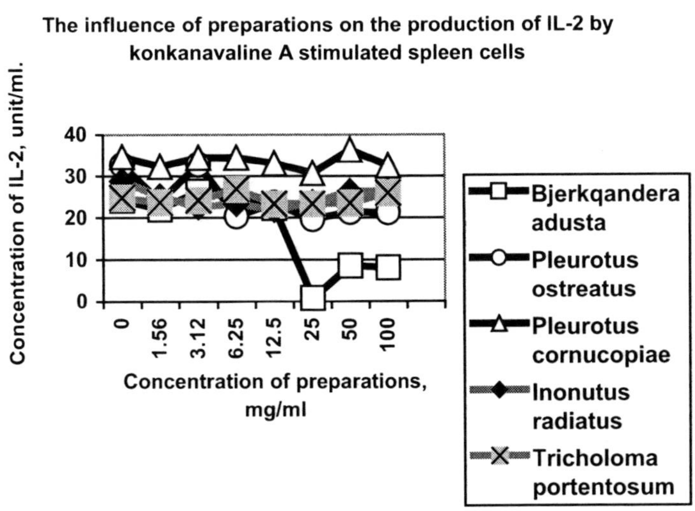

Figure 6.

Thus, the studied mushroom extracts obtained as from natural (so called wild) fruit bodies, as from biotechnological (submerged mycelia) samples possess expressed immunomodmulating action, involving in response of various immunocompetent cells. We marked mitogenic effect of mushroom extracts on lymphoid cells, stimulating action on monocytic cells of a peripheral blood synthesizing proinflammatory cytokines, and also ability of these preparations to promote main cells of inflammatory response - neutrophils, causing not only chemotaxis but also oxygen burst showing an activation of phagocytic response. Taking into account obtained data, we can predict, that studied preparations generally are causing direct stimulating effect on B-lymphocytes and macrophages, and significantly weaker influencing functional activity of T-lymphocytes.

At the same time it is clear, that the extracts of diferent mushrooms vary in stimulating of different immunocompetent cells.

The outcomes obtained in the experiments *in vitro* have predetermined check of these extracts on antitumor effect in experiment *in vivo*. As a model of tumor growth one of the most malignant tumours melanoma B16 was selected.

The experiments have been done using 16 C57BL female mices (average body weight 23 g) purchased from animal facility "Rappolovo", Russian Academy of Science. Animals were kept 8 per one standard polypropylene chamber. The light regimen was standardized to 12 hours light/dark cycle. All animals received water ad libitum. Animal were divided into two groups – control and experimental. The control group was given uzual fodder. The second group animals were given modified fodder with extract (daily dosage 100 mg kg^{-1} body weight in recalculation on a dry extract).

During two week periods from a beginning of extract administration the average body weight did not differ significantly as compared to control.

Table 5. The effect of *Pleurotus ostreatus* extract peroral intake on tumour development in mices bearing melanoma B16

Time after tumour transplantation, days	Tumour size, sm^3		Suppression, %
	Experimental group	Control group	
8	0.02 ± 0.02	0.03 ± 0.01	30
12	0.14 ± 0.09	0.43 ± 0.18	70
15	0.55 ± 0.33	2.19 ± 0.45	75
22	1.52 ± 0.52	4.34 ± 1.24	65
26	1.84 ± 1.30	9.37 ± 2.64	80
29	0.28 ± 0.17	-	-
33	0.11 ± 0.02	-	-
36	0.34 ± 0.01	-	-
40	0.67 ± 0.02	-	-
43	1.22	-	-

At 15 day melanoma B16 was transplanted to all animals (10^6 cells of a melanoma B16 subcutaneously to the right abdomen side). In further measurement of parameters of tumour growth and life duration was conducted.

The progressive and irregular decrease of average body weight was noted in both groups after tumour transplantation but less in group received extract. At 26 day after tumour transplantation average body weight was significantly higher in experimental group.

The average tumour size changes are represented in table 5. In control group average tumour size increased during tumour development. In experimental group average tumour size decrease 30-80 % was marked with maximal effect during first 14 days.

The survival rate of mice with melanoma B16 is represented in a fig.7.

Figure 7. The survival of mice bearing melanoma B-16

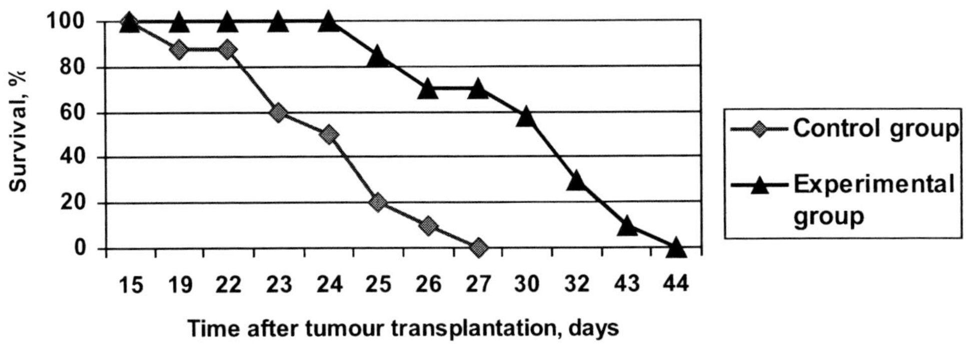

The average and maximal duration of life in check group reliably is lower in comparison with experimental group.

Thus, the extract from *P. ostreatus* fruit bodies makes reliable inhibiting effect on melanoma B-16 growth and decreases a manifestation of tumorous intoxication as it confirms by the weight differences between control and feed with mushroom extract mice. These effects perhaps are connected with the stimulation of immune system which has a significant importance for a pathogenesis of the tumour growth.

All of the data obtained during our investigations suggest that aqueous extracts of studied Basidiomycetes have marked biological effects. However further investigations on the influence of this extracts on the immune system are necessary, since the specter of their influence is possibly significantly wider and diverse.

References

[1] F.J. Gregory, E.M. Healy, H.P.K., Jr. Agersborg, G.H. Warren: Studies of antitumor substances produced by Basidiomycetes. *Mycologia,* 58, 80-90 (1966)

[2] M.A. Espenshade, E.W. Griffith: Tumor-inhibiting basidiomycetes. Isolation and cultivation in the laboratory. *Mycologia,* 58, 511-517 (1966)

[3] T. Ikekawa, N. Uehara, Y. Maeda, M. Nakanishi, F.Fukuoka: Antitumor activity of aqueous extracts of edible mushrooms. *Cancer Res*, 29, 734-735 (1969)

[4] T. Mizuno The extraction and development of antitumor-active polysaccharides from medicinal mushrooms in Japan (review). *Int J Med Mushr,* 1, 9 – 29 (1999)

[5] S.P. Wasser, A. Weis: Medicinal properties of substances occuring in higher basidiomycetes mushrooms: current perspectives (review). *Int J Med Mushr*, 1, 31-62 (1999)

[6] S.T. Chang: Global impact of edible and medicinal mushrooms on human welfare in the 21st century: nongreen revolution. *Int J Med Mushr*, 1, 1-7 (1999)

[7] A. Boyum: Isolation of mononuclear cells and granulocytes from human blood. Isolation of mononuclear cells by centrifugation. *Scand J Clinic Labor Invest,* 41, 77 – 89 (1968)

[8] M.M. Dike, J.K. Forlin (Eds): *Manual of Immunopharmacology*. Moscow, 1998. 332 p (in Russian).

[9] T. Tono-Oro, N. Ueno, T. Matsumoto: Chemiluminescence of Whole Blood, a simple and rapid method for the testimation of phagocytic function of the granulocytes and opsonic activity in Whole Blood. *Clin Immun Immunopath*, 26, 66 – 75 (1990)

In: Biotechnology in Agriculture and the Food Industry ISBN 1-59454-119-1
Editor: G. E. Zaikov, pp. 199-203

Chapter 27

SAFETY ASSESSMENT OF PROTEIN CONCENTRATE DERIVED FROM GENETICALLY MODIFIED SOYBEAN, LINE 40-3-2: EFFECT ON SOME BIOCHEMICAL PARAMETERS

V. A. Tuteljan, L. V. Kravchenko E. Yu. Sorokina, N. V. Lashneva, L. I. Avrenieva, G. V. Guseva, and O. N. Chernyshova

Institute of Nutrition, Russian Academy of Medical Sciences
109240, Moscow, Russia, Ustinski Proezd, 2/14.
Tel/fax: 298-18-66
E-mail: Sorokina@ion.ru

ABSTRACT

The rats were fed 1.25 g/rat/day of protein concentrate derived from genetically modified soybeans, line 40-3-2 (Monsanto Co., USA) with diet for 5 months. Their blood checked for total protein and glucose level, aminotransferase and alkaline phosphatase activity; urea for pH, relative density and creatinine level, and liver for enzyme activity of the phase I and II of xenobiotic metabolism, and the total and non-sedimentable lysosomal enzyme activity. It has been shown that when rats receive the protein concentrate derived from genetically modified soybeans in their diet for a long time, there were no changes in cellular membrane function and enzymatic activity beyond physiologic

Key words: Genetically modified soybeans, Glyphosate tolerant soy, rats, protein concentrate

Recently, some countries produce food or food components by means of genetic engineering. Using this technology we can develop highly productive, resistant to diseases and pests crops and foods with improved nutrition value and taste. Today there are more than 80 transgenic crops exist, such as wheat, rice, corn, tomato, potato, soybeans, etc. Of these crops, genetically modified soy, mainly glyphosate resistant, holds the first place in production and consumption [1].

In plants, glyphosate inhibits 5-enolpyruvyl shikimate-3-phosphorate synthase (EPSPS) which is the key enzyme for synthesis of aromatic aminoacids, which leads to suppression of growth and eventually to death of the plant. Glyphosate tolerant soy was developed by methods of genetic engineering and gene isolated from *Agrobacterium sp.* tolerant to glyphosate [2]. Products derived from genetically modified soy do not differ from those made of conventional soy with regard to chemical composition and technological parameters [3]. Experiments *in vitro* and *in vivo* have proved that the protein of glyphosate tolerance synthesized as a result of genetic modification is readily dissipated in the gastrointestinal tract of mammals and is not toxic to laboratory animals [4].

Genetic engineering is an extremely promising technology; however there is a certain probability of an unwanted effect which might arise due to unexpected gene expression. From medical point of view, the concern is a probable decrease in nutrition value, or altered amount of ordinary metabolites, or new toxic minor components which might appear as a result of changes in the genetically modified plant metabolism [5, 6].

On the basis of our own studies and literature data, we believe that the activity of enzymes involved in the protection and adaptation process is one of the mostly informative parameters for the assessment of genetically modified food. These are xenobiotics metabolizing enzymes which are highly and non-specifically sensitive to any toxic and alimentary factors and enzymes of lysosomes which are cell organelles involved into various pathological processes in the early stages [7-10].

Objective of this study was to assess the effect of protein concentrate (ADM, USA) derived from genetically modified soy, line 40-3-2 (Monsanto Co., USA) tolerant to glyphosate, on some biochemical parameters in rats. Protein concentrate derived from conventional soy (ADM, USA) was used as control. Wistar male rats with the initial body weight of 72-74 g were divided into three groups Group 1 (control-1) fed the standard diet during the whole period of the experiment (5 months) Group 2 (control-2) fed a standard diet + 1.25 g of protein concentrate from conventional soy (20% of all protein). Group 3 (experiment) fed the standard diet + 1.25 g of protein concentrate from GM soy.

Blood biochemical parameters were determined one and five months from the beginning of the experiment. Blood serum was tested for total protein - by the standard method: biuret reaction, stain registration at 540 nm; glucose content - by glucose oxidant method; activity of alanine aminotranspherase and asparte atminotranspherase - by standardized method for optimized optical test; activity of alkaline phosphatase; level of creatinine in urea, stain registration at 505 nm [11]. Total and non-sedimentable activity of lysosomal enzymes: arylsulphatase A and B, β -galactosidase, β-glucoronidase, were determined in liver homogenate and cytosole fractions [12]. Cytochrome *P-450* and b_5 content [13], the rate of *O*-deethylation of 7-ethoxycoumarine and *N*-demethylation of aminopyrine [14], the activity of acetylesterase [15], epoxidhydrolase [16], UDP glucuronosiltranspherase with *p*-nitrophenyl as an acceptor [17] in microsomes isolated from rat liver were detected.

As has shown in Table 1, the level of total protein and glucose, activity of aminotranspherasa and alkaline phosphatase in the blood serum in animals of group 1 is statistically confident, and do not differ from these parameters of the control group during the whole period of the experiment. Biochemical analysis shows urea pH 6.9-7.0 in all groups, and daily diuresis. Urea relative density and creatinin content in rats of the experimental group was the same as those parameters in the control group which is within the physiological norms for rats.

Table 1. Blood biochemical parameters in rats received protein concentrate derived from genetically modified and conventional soy*

Parameter	Timing, months	Group of animals (*n=8*)		
		Control-1	Control-2	Experiment
Total protein, g/l	1	64.2±1.8	65.2±2.9	64.8±3.3
	5	82.4±3.7	75.2±4.9	76.8±2.2
Glucose, mmol/l	1	8.6±0.3	9.0±0.5	9.7±0.5
	5	6.8±0.3	6.9±0.4	6.9±0.3
Alanine aminotranspherase, µmol/min l	1	54.4±5.4	66.1±6.4	56.0±4.8
	5	56.5±6.3	61.8±3.5	65.8±6.0
Aspartate aminotranspherase, µmol /min l	1	60.0±6.1	67.5±4.4	61.0±3.4
	5	48.9±3.4	66.0±3.8	47.7±4.7
Alkaline phosphatase, unit/l	1	805.5±89.3	722.2±65.3	822.5±36.5
	5	809.9±60.3	522.2±45.7	529.5±38.3

*Values are means ± SEM. No significant differences were observed (P>0,05).

Studies of total and non-sedimentable activity of lysosomal enzymes in rats show that the diet with protein concentrate derived from genetically modified soy given to rats during one to five months doesn't affect function of lysosomal system in rat liver (Table 2, 3).

Table 2. Total activity of hepatic lysosomic enzymes in rats who fed protein concentrate derived from genetically modified and conventional soy*

Enzyme, µmol/min g of tissue	Timing, months	Group of animals (*n=6*)		
		Control-1	Control-2	Experiment
Arylsulphatase A and B	1	2.49±0.04	2.48±0.04	2.43±0.04
	5	2.43±0.08	2.56±0.11	2.62±0.12
β -galactosidase	1	2.14±0.04	2.16±0.04	2.17±0.04
	5	2.72±0.09	2.79±0.07	2.65±0.04
β -glucoronidase	1	2.45±0.06	2.57±0.07	2.49±0.05
	5	2.35±0.12	2.38±0.09	2.48±0.14

*Values are means ± SEM. No significant differences were observed (P>0,05).

Studies of hepatic microsomal enzymes in rats (Table 4) show that by the end of the 1st month of the experiment, it was a mild statistically significant increase in the amount of cytochrome *P-450* and b_5, and in the activity of 7-ethoxycoumarin deethylase, to the same extend in rats received genetically modified and conventional soy. Apparently, such changes are related to protein concentrate in rat's diet regardless of the source of the protein. By the end of the experiment, no significant differences was registered in animals of various groups in terms of the activity of xenobiotic metabolism enzymes, content of cytochrome *P-450* and b_5, and microsome protein in rat's liver (Table 4).

Table 3. Non-sedimentable activity of hepatic lysosomic enzymes in rats who fed protein concentrate derived from genetically modified and conventional soy*

Enzyme, % of total concentration	Timing, months	Group of animals (*n=6)*		
		Control-1	Control-2	Experiment
Arylsulphatase A and B	1	3.30±0.13	3.23±0.11	3.22±0.12
	5	3.43±0.15	3.53±0.09	3.30±0.12
ß-galactosidase	1	5.32±0.32	5.26±0.14	5.18±0.13
	5	5.66±0.34	5.80±0.49	5.88±0.28
ß -glucoronidase	1	6.51±0.32	5.86±0.21	6.56±0.29
	5	4.90±0.34	4.90±0.11	4.91±0.15

*Values are means ± SEM. No significant differences were observed (P>0,05).

Table 4. Activity of xenobiotic metabolizing enzymes and protein content in hepatic microsomes in rats fed with protein concentrate derived from genetically modified and conventional soy

Parameter	Timing, months	Group of animals (*n=6)*		
		Control-1	Control-2	Experiment
Cytochrome *P-450, nmol/mg* of protein	1 5	0.50±0.07 0.60±0.08	0.72±0.06* 0.58±0.05	0.71±0.02* 0.61±0.03
Cytochrome b_5*, nmol/mg* of protein	1 5	0.48±0.04 0.59±0.05	0.66±0.05* 0.56±0.09	0.66±0.02* 0.67±0.07
O-deethylation of 7-ethoxycoumarin, *nmol/min mg of* protein	1 5	2.0±0.46 1.68±0.23	3.19±0.52* 1.42±0.15	3.52±0.25* 1.33±0.23
N-demethylation of aminopyrine *nmol/min mg of* protein	1 5	10.37±0.21 10.12±0.29	10.45±0.44 10.57±0.37	10.13±0.38 11.13±0.18
Acetylesterase, µmol/*min mg of* protein	1 5	7.44±0.32 6.05±0.32	7.77±0.24 5.58±0.33	7.74±0.30 5.03±0.60
Epoxidhydrolase, *nmol/min mg of* protein	1 5	6.00±0.14 5.55±0.14	5.46±0.11 5.05±0.53	5.54±1.40 4.87±0.47
UDP-glucuronosiltranspherase, *nmol/min mg of* protein	1 5	20.8± 3.3 18.1± 0.4	23.1± 1.5 16.1± 2.1	22.4± 2.2 15.9± 1.7
Protein, mg/g	1 5	14.2± 0.96 14.6± 0.50	16.1± 0.92 16.0± 0.40	15.3± 0.60 15.9± 0.60

* - statistically significant difference in parameters, *p<0.05, as compared with control-1*

Thus, it has been shown that when the rats receive protein concentrate derived from conventional or genetically modified soy with diet for 5 months, it does not have any effect on studied biochemical parameters of blood and urea in rats, does not cause any changes in the activity of the enzymes of phase I and II of xenobiotics metabolism, localized in membranes of endoplasmatic reticulum, does not affect the activity of lisosomal enzymes, and it does not cause changes in lisosomal membrane stability.

On the basis of the studies conducted, we conclude that if rats receive protein concentrate derived from genetically modified glyphosate tolerant soy with diet for a long period of time, it does not cause any significant changes beyond the physiological limits in function of cellular membranes, and the activity of enzyme systems involved in biotransformation and detoxification of endo- and exsogenic toxic compounds.

The results obtained supplement and confirm conclusions made on the basis of previous studies on safety assessment of glyphosate tolerant soy line 40-3-2 [18]. This, subchronic studies (conducted during one month) in caws, chickens, fish, rats and mice proved good tolerance to fodder containing modified soy, line 40-3-2 and lack of any toxic effect. This allowed the authors to conclude that genetically modified soy line 4-3-2, is identical to the conventional soy in terms of the composition, nutrition value and food safety.

REFERENCES

[1] C. James: *Global Status of Commercialized Transgenic Crops*: ISAAA Briefs,27,2002 24 p

[2] S.R. Padgette, K.H Kolacz, Y.X.Delanna et al. *Crop Sci,* 35, 1451-1461 (1995)

[3] S.R Padgette., N.B Taylor., D.L Nida. et al. *J Nutr,* 126, 702-716 (1996).

[4] L.A Harrison., M.R. Barley, M.W. Naylor et al. *J Nutr*, 126, 728-740 (1996).

[5] WHO. Strategies for assessing the safety of food produced by biotechnology. *Report of a Joint FAO/WHO Consultation.* WHO, Geneva, 1996.

[6] WHO. *Environmental Health Criteria 6. Principles and Methods for Evaluating the Toxicity of Chemicals.* Part 1. WHO. Geneva, 1978

[7] A.A. Pokrovsky, V.A. Tuteljan *Lysosomes.* Moscow, Nauka, 1976, 234 p (in Russian)

[8] V.A. Tuteljan, L.V. Kravchenko, E.E. Kuzmina *Drug Metabolism and Disposition*, 1, 223-225, (1990).

[9] V.A. Tuteljan, L.V. Kravchenko, L.I. Avrenyeva, E.E. Kuzmina. *Nutr Research*, 10, 1119-1129 (1990)

[10] L.V. Kravchenko, E.E. Kuzmina, L.I. Avrenyeva et al. *Vopr. Pitania*, 4, 13-16 (1994) (in Russian)

[11] Reference book. *Laboratory Methods to Conduct Sturies in a Clinic.* Edited by V.V. Menshikov. M; Medicine 1987; 174-175, (in Russian).

[12] G. Dingl. *The Lysosomes. Methods of study.* Moscow, 1980, 350 p (in Russian)

[13] T. Omura, R.J. Sato. *Biol Chem*, 239, 2370-2377 (1964).

[14] T.Nash. *Biochem J*, 5, 416-421 (1953).

[15] A.A. Pokrovsky, A.I. Archakov. *Reports of the USSR Academy of Sciences.*, Moscow, Nauka 1976 (in Russian)

[16] L.V. Kravchenko, E.E. Kuzmina, V.S. Sobolev *Vopr. Med. Chemistry* 6, 122-123 (1988) (in Russian)

[17] V. Burchell, P. Weatherill *Methods in Enzymology*, 7, 169-176 (1981).

[18] B.G. Hammord, J.L. Vicini, G.F. Hartnell et al. *J Nutr*, 126, 717-727 (1996).

In: Biotechnology in Agriculture and the Food Industry ISBN 1-59454-119-1
Editor: G. E. Zaikov, pp. 205-208

Chapter 28

BIOTECHNOLOGY OF PROBIOTIC FOODS BASED ON MILK AND RAW PLANT MATERIALS

L. V. Krasnikova and L. A. Zabodalova
Saint-Petersburg State University of Refrigeration and Food Engineering
191002, 9 Lomonosov St., Saint Petersburg
E-mail: refr @ sarft.spb.ru, fax (812) 315-37-78

ABSTRACT

The purpose of the study was to develop fermented beverages biotechnologically on the base of the whole and non-fat milk and milk whey. Fermentation of the initial mixture was achieved by combining probiotic cultures Lactobacillus acidophilus, Lactobacillus casei subsp.rhamnosus, Bifidobacterium bifidum in certain proportions. Cultures included in the combination were selected according to their high antagonistic and enzymic activity. The sour base was filled with different additives of plant origin.

Key words: Probiotic foods, normal intestinal microflora, whole and non-fat milk, milk whey, black currant puree, fruit powders.

Human ecological safety has become one of the main geopolitical issues by the end of the 20th and at the turn of the 21st century. Severe ecological impacts of scientific-technological advances have drastically changed the human habitat.

Preservation of human health comes to be of primary importance as adverse impacts from outside are growing in number, such as ecologically unfavourable living and harmful working conditions, repeated stress situations, including chronic tiredness syndrome, high ionizing and other radiation background, extensive application of antibiotics and chemotherapy, etc. It is essential to create such quality of life that will allow to neutralize pathogenic potential of the environment. Usage of medicated ways to correct homeostasis often produces undesirable side-effects on organism. Therefore, improvement of homeostasis as well as prevention of its disorders by means of foods of new quality seems promising, these foods providing energy beyond basic nutrition, enhancing immunity and adapting the consumers to the environment.

When systematicalli used such foods render some regulating action on the organism as a whole or selectively on its separate systems and organs. Such foods are called functional.

Most common and mass functional foods are, first of all, those which favour improvement and sustention of microbial ecology of the humans and, in the first place, microflora of gastrointestinal tract. Depending on the manner of recovering the microflora of the host one can differentiate prebiotic, probiotic and synbiotic foods [1].

Probiotics contain live cells of microorganisms or food additives of microbial origin. Their beneficial effect on the organism is shown due to regulation of the gut microflora.

Prebiotic foods involve prebiotics, i.e. substances that stimulate growth and activity of normal intestinal microflora. However, maximum effect is known to be achieved when probiotics and prebiotics are rationally combined. Such products are referred as synbiotics [2].

The present investigation was aimed at developing biotechnology of fermented milk beverages on the basis of whole and non-fat milk and milk whey. The original mixture was fermented with the combination of probiotic cultures L. acidophilus, L.casei subsp. rhamnosus, B. bifidum conformably correlated.

Bearing in mind the concept of medicinal-prophylactic beverages, the most important quality of lacto- and bifidobacteria is their ability to reduce the growth of pathogenic and conventionally pathogenic microorganisms [3]. Table 1 represents the data on the antagonistic activities of bacteria species selected as a starter for the fermented milk beverages.

Table 1 Antagonistic activities of lacto- and bifidobacteria

Test-organisms	Inhibition zones of test- organisms growth		
	L.acidophilus 20T	L.casei subsp. rhamnosus C25	B.bifidum КС-7
Shigella sonnei 865	17±2	8±1	28±3
Shigella flexneri 331	16±2	7±1	28±3
E. coli 0144	11±1	6±1	15±2
Salmonella enteritidis 36	11±1	7±1	28±3
Proteus mirabilis 165	10±1	6±1	17±2
S. aureus 209	10±1	Отсут.	4±1

Bifidobacteria strain was found to possess the highest antagonistic activities against the majority of pathogenic bacteria. L. casei subsp. rhamnosus strain was noted to be inferior to the L. acidophilus and bifidobacteria in this factor.

To extend the assortment of beverages proposed and to impart them with the medicinal-prophylactic qualities against GI-tract diseases, various filling materials of plant origin were used. Among them were fruit purees and dry powders, syrups made of Fructus rosae, Silybum marianum, Hippophaes rhamnoides, as well as natural colour of Aronia melanocarpa, that gave the ready-to-drink beverage desirable colour and special flavour.

Both syrups and purees made of black currant (Ribes Negrum) were administred during the final stage of ripening, the quantities being determined experimentally.

From a number of widely known fruit powders apple one was preferred as readily available and having high content of pectin substances and fibers (ca. 14 % and 20 % on solids respectively). Side by side with this, red billberry powder was used.

Fruit powders were put into the fermented basis in the form of puree. Being finely dispersed media, the powders absorb some quantity of water when they are wetted. It is possible to observe swelling of high molecular components of the powder, pectin matters in particular. The process of swelling, to a considerable extent, depends on the ratio of powder:water (water duty), temperature and the time of hydration. Maximum rehydration of powders – defined by the amount of water bound by 1 g. – was observed with water duty 1:10. However, such water duty value causes substantial dilution of the product, which results in lowered content of dry solubles, including biologically active matters. In the later investigations water duty value 1:5 was chosen as it was taken into account that lacking water can be absorbed by the powder from the fermented milk beverage. The investigation showed that dry soluble matter of the powder was rehydrated intensively during the first 15-20 minutes of its contact with water. Solubility of the powder was noticed to rise to some extent when the temperature of water ranged from 15 to 45 °C. However, to avoid unreasonable wasting of power it was practical to limit the temperature range to 20-25 °C. Thus the following parameters of powder preparation were adopted: water duty value – 1:5, water temperature – 20-25 °C, time of hydration – 15-20 min. Under these conditions the schedule of making apple puree from dry powder looks like following sequence of operations:

Raw material handling → Powder rehydration (water duty 1:5, water temperature – 20-25 °C) → Hydration (τ = 15-20 min) → Pasteurization (temperature 76 ±2 °C, time of holding 20 s.) → Cooling (down to 30-35 °C) → Administration of puree to the product (15 % by mass).

When the vitamin composition of the beverages under consideration was studied, it became clear that administration of the above filling matters enriched the product with vitamins, especially with ascorbic acid and niacin, their contens growing 4,0 - 4,5 and 2 times respectively (Table 2).

Table 2. Vitamin contents in probiotic beverages

Product	Vitamins, mg/100 g product				
	B_1	B_2	PP	C	β-carotin
Fermented milk beverage without filling matter	0,030	0,105	0,11	0,82	0,01
Fermented milk beverage +15% black currant puree	0,030	0,152	0,19	3,86	0,01
Milk whey	0,025	0,100	0,12	0,62	trages
Whey beverage + 3 % Fructus Rosae syrup + 1 % Hippophaes rhamnoides syrup	0,027	0,138	0,21	5,98	1,16
Whey beverage + 3% Fructus Rosae syrup + 1 % Silybum marianum syrup	0,027	0,138	0,22	4,21	0,15
Whey beverage + 2 % Aronia melanocarpa color concentrate	0,018	0,140	0,18	0,98	0,18

It should be noted that β-carotin content considerably increased in the products enriched with syrups of Fructus Rosae, Hippophaes rhamnoides, Silybum marianum and color concentrate from Aronia melanocarpa.

The results of clinical testing allow to validate the efficiency of the developed products to be used for treating GI-tract diseases of various aetiology.

References

[1] B.A. Shenderov *Medical microbial ecology and functional nutrition. Vol. III. Probiotics and functional nutrition*. GRANT, Moskow, 2001. 288 p. (in Russian)

[2] R.D. Berg: *Probiotics, prebiotics or "conbiotics". Trends in Microbiology* (United Kingdom), 6 (3) 89 (1998)

[3] M. Aguirre, M.D. Collins: Lactic acid bacteria and human clinical infection. *J. Appl. Bacteriol.* 75 (2) 95 (1993)

In: Biotechnology in Agriculture and the Food Industry
Editor: G. E. Zaikov, pp. 209-213
ISBN 1-59454-119-1

Chapter 29

MORPHOLOGICAL STUDY OF THE BASIDIOSPORES GERMINATION OF *BOLETUS EDULIS*

G. G. Zharikova and M. V. Nefelova
Plekhanov Academy of Economics,
Laboratory of Microbiology and Mycology.
E-mail svgk@REA.ru
phone: 237-94-98
fax: 237-87-65
Moscow 113054 Stremyanny per. 36

ABSTRACT

The possible commercialization of a cepe, dependent on methods of its coltivation, demands detailed and comprehensive study of the development cycle *Boletus edulis,* basidiospores germinating of initial mycelium growing in artifical conditions. The new method of basidiospores incubation allows to observe morphological changes and to show, that at an incubation in fluid solutions the changed deformed spores are aggregated in dense globula or plasmodies. This new heterogenous biological clusters germinating of young vegetative mycelium. Growing of a colony - spawn of mycelium have legible space and time organization, contain identical building blocks with different

Key words: *Boletus edulis,* basidiospores, mycelium, morphology.

Cultivation of maximum edible mushrooms and the obtaining in artifical conditions of fruit bodies or biomass of mycelium represents definite practical value. This is developed for several and prolongs to be developed for series species of mushrooms /1, 2/. In this respect the major interest represents *B. edulis,* having high alimentary, delicious-flavor and medical qualities /3, 4/.

There are no known reports of cultivation of *B. edulis.* In the literature intercommunicates about major difficulty of germinating basidiospores in artifical conditions and about possibility of obtaining of growing vegetative mycelium from slices of young fruit bodies of a fungus /5, 1/.

The problem of cultivation of a cepe demands deep and comprehensive study of morphology, cytology, physiology, genetics of ectomycorrhisal operation of the object, that can get a complete alternation of development of a fungus *Boletus edulis*. The beginning of a cycle is the process of a basidiospores germination and grow of initial mycelium. The study of a morphgenesis of the called processes has constituted a problem of the given research.

Basidiospores were obtained from basidiocarps of Boletus edulis collected in 1999, 2000, 2001 in forests of Russia. The spores in an amount 10^4 - 10^5 were introduced in 0,1 - 1 % sterile solutions of organic matters /6, 7/. The incubation was conducted at the temperature of 18 - 20 °C during 20 - 30 day. Of different age the assays for microscope examination were taken. The growing colonies of mycelium were carried on microscope slides without change of the form of colonies. The native sample of spores and colonies of mycelium researched in a light microscope.

Under study of germinating of fungus spores usually the grow of initial tubules or grow colonies on agar plate is noted. This is important that in given work dynamics of the process from beginning of a germination, before the formed colonies of vegetative mycelium was tracked.

The work was conducted with cultures, in which the stranger microflora (bacteria or lowest fungi) did not educe, that was controlled by special experiences.

During incubation in stationary fluid culture basidiospores swalled, are deformed and to 5 - 6 day the modified spores and remainders of spores of a different degree of deformation drew together and formed in compact globula /Fig. 1/.

At effect of a thermal shock spores expose even more to radical changes. There is a destruction of the cell wall of spores, transformation in the spherical protoplasts, surrounded by a membrane, which merge in larger formless plasmodia /Fig. 1/. To 15th - 20th day of incubation on a surface of clumps of the changed spores the germ and hypha appeared, which at a prolonging incubation expand, branch, derivating the network of vegetative mycelium /Fig. 1/.

In a clump thus of spores, quite probably, there are functional variations of spore, remainders and parts of spores, one of which shape germs, and other expose apoptose and will be used as a nutrient stuff for growing mycelium.

Besides the clump of deformed spores, most likely, contains and generically variable heterosexual remainders of spores. In a development cycle of basidiomycetes from separate heterosexual haploid spores the initial haploid mycelium grows, after hyphal fusion which educes already dikaryotic mycelium /8, 9/. At growing colony - mycelium from a genetically heterogenous clump deformed basidiospores the coalescence at the stage of the initial haploid cells and direct growing of dikaryotic mycelium is possible, that demands special study.

Growing in liquid stationary culture the volumetric colony of vegetative mycelium *Boletus edulis* by 30 day of cultivation achieves 25-35 mm of height and 10-15 mm of width. At some variations of an exterior form of colony always have identical and legible organization of single structure and definite architectonics.

The basis of the colony is formed by a dense globule of glyed modified basidiospores and their remainders, which has a dark brown colour. Morphology of the globule changes as the colony grows: the structure becomes more homogeneous and less dense, the size of globulae decreases, in grown colonies it can disappear. From the globulae escape thick draws-hoses, they have dark content, are not separate and can be compared to conductive of feeding draws described with other fungi (Kursanov, 1940, Cherepanova, 1981). Some long draws protrude to the colony centre, parts of draws can be seen in different sections of colonies /Fig.2/.

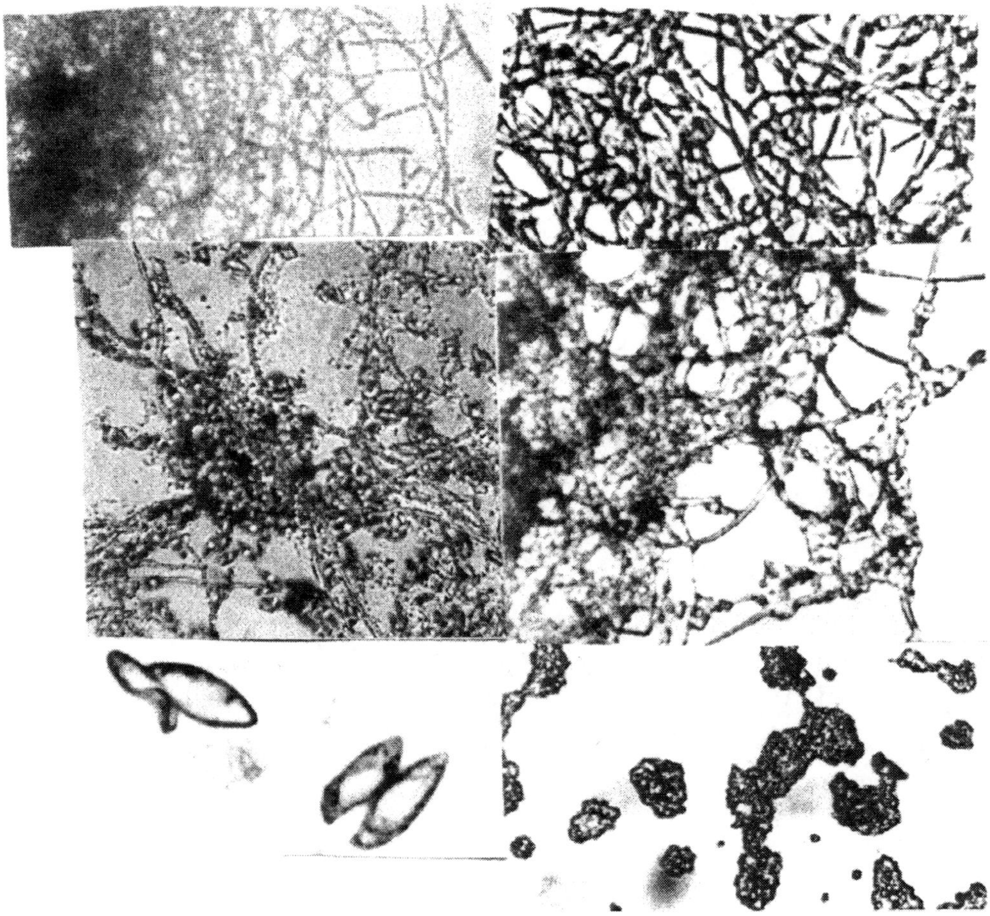

Fig. 1. Morphgenesis of Boletus edulis basidiospores during an incubation in liquid stationary culture: a - swelling and deformation of spores, 900 x; б - clump of spores in globulae, 200 x; в - growth of vegetative mycelium, 200 x; г - protoplasts and plasmodium after effect of a thermal shock, 400 x; д - derivation a hypha and cords, 400 x; e - growth of vegetative mycelium, 400 x.

The principle mass of the colony-spawn is formed by a network of vegetative mycelium. Mycelium hypha are straight, separate with the diameter of 5 - 7 um. There are anastomoses, cords of 4 - 5 parallel hypha, clasps were not found. Walls of hypha in the centre of the colony are covered with a layer of gel. Ends of young growing hypha are surrounded with drops of gel secretion /Fig.2/.

On all vorum of colony there are also dense spherical clumps of closed-grained mass. This substance has not an external membrane and has darker colour than mycelium.

The whole colony-spawn is enveloped by the shell sometimes of multilayer mycous mass.

The heterogeneity, functional differences of a basidiospore biomass, from which the colony grows, legible spatio-temporal organization of a colony-spawn -mycelium *Boletus edulis* will be agreeed the modern concepts about a role of differenciation of cells and formation of component in colonial assemblage of microbial populations /10/.

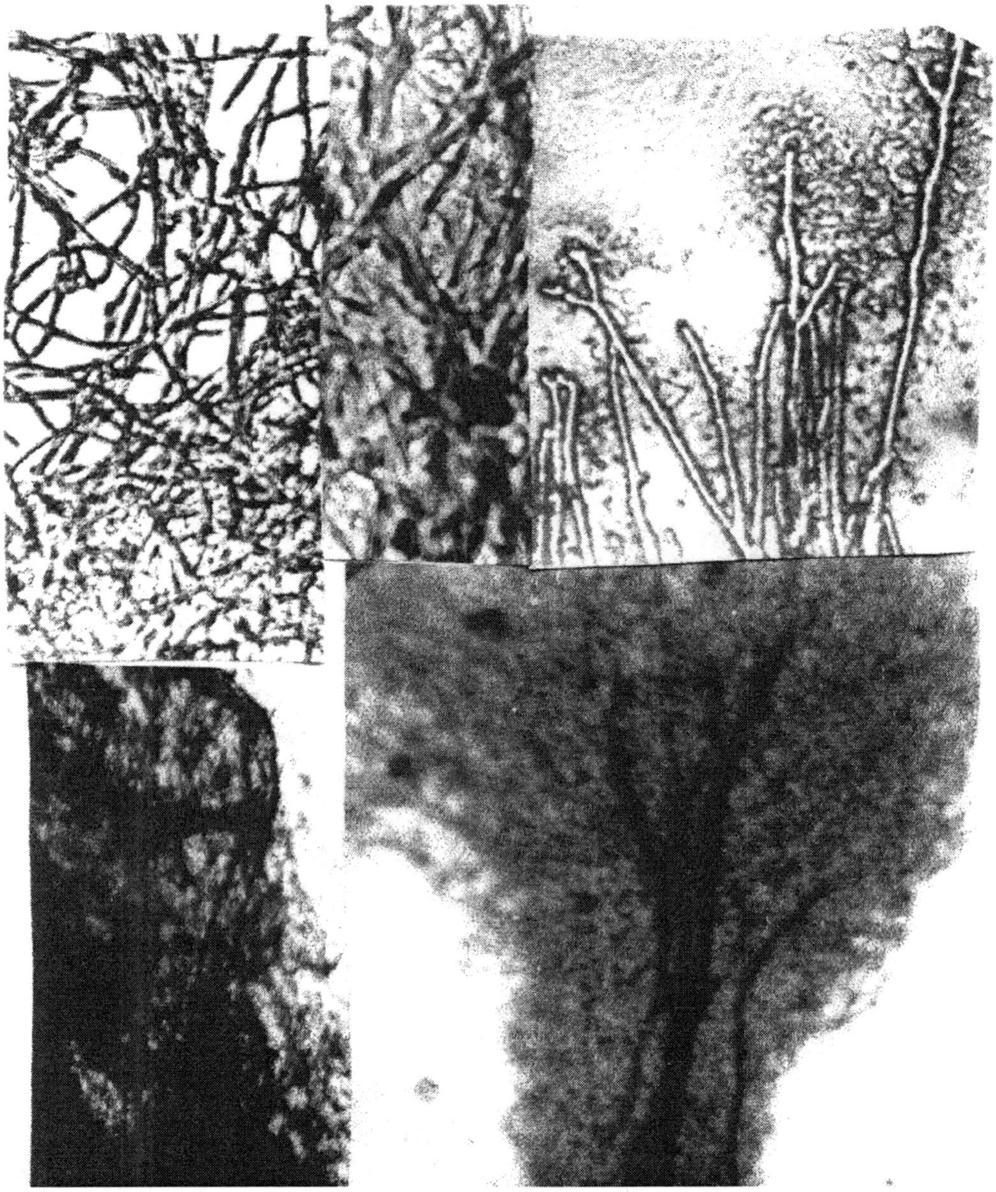

Fig. 2. Structure of parts of *Boletus edulis* colony in liquid stationary culture: a - globulae of spores and growing hypha - drows, 200 x; б - dense network of growing mycelium, drows-hoses, 100 x, grainy, dark particles; в - swelling mycelium of the center of colony, 400 x; г - peripheric part of colony, young mycelium, 400 x; д - apical ends of hypha coated by gel drops, 400 x.

Growing in artifical conditions from basidiospore mass *Boletus edulis* the colony of vegetative mycelium is the valuable object for further studying as theoretical problems, bound with a development cycle of a fungus, and solution of practical problems of obtaining of vegetative mycelium biomass.

REFERENCES

[1] N.A.Dudka (ed.): *Industrial cultivation of edible fungi*. Kiev, 1978.

[2] K.E.Mshigeni, Shu-Ting-Chang (ed's): *Mushrooms: The biology, nutritional and medical properties, cultivation, technologies*. Namibia, 2001.

[3] A.V.Shubin: *Boletus edulis. A bionomics and fructification*. Petrozavodsk, 1986.

[4] J.R.Hall, A.J.Lyon, J.Wang, L.Sinclair: *Economic Botany*. 1998, 52 (1), 44.

[5] O.Modes: *Symble Botanical Upsalienses*. 1941, 1, 5.

[6] V.I.Vidyapin, G.G.Zharikova, M.V.Nefelova, E.N.Zharikova, V.M.Stasevitch: *The patent № 2188536*. Russia, 2002.

[7] V.I.Vidyapin, G.G.Zharikova, M.V.Nefelova, E.N.Zharikova, A.O.Kozmina, V.M.Stasevitch: *The patent № 2188537*. Russia, 2002.

[8] E. Moore - Lindecker: *Fundamental of The Fungi*. L, 1982.

[9] Y.T.D'yakov: *Introduction to an algology and mycology*. Moscow, MSU, 2000.

[10] A.V.Oleskin, I.V.Botvinko, E.A.Tsevkelova: *Microbiology*, 2000, 69 (3), 309.

In: Biotechnology in Agriculture and the Food Industry
Editor: G. E. Zaikov, pp. 215-222
ISBN 1-59454-119-1

Chapter 30

OPTIMIZATION OF THE MIXED ALBUMINOUS – CARBOHYDRATED FOODSTUFF FORMULATION

***V. E. Zhidkov*[1]*, V. V. Sadovoy*[2] *and A. A. Brykalova*[1]**

[1]Stavropol State Technological Institute of Economics and Service, the Branch of the Southern–Russian State University of Economics and Service.
76/6, Shpakovskaya St., Stavropol, 355037, Russia
E-mail: stis @ statel. Stavropol. ru, факс 77-99-63
[2] The Southern Caucasian State Technical University
2, Kulakov Ave., StavropoL, 355029, Russia
E-mail: sstu @ stv. runnet.ru, факс 95-68-08

ABSTRACT

The methods of designing mixed formulations based on soy–bean concentrate and milk albuminous – carbohydrates components are described. For the purpose of improving the functionally– technological properties of foodstuffs the optimization of the compositional mixture has been carried out by means of the "Neuron Technique". The variants of formulas possessing a high capability to form gels and emulsions to bind water and fat have been established.

Key words: soy–bean concentrate, nonfat dry milk, lactose-protein complex (lactobal), dry whey and casein-albuminous complex.

Nowadays different food admixtures are widely used in the food industry in order to improve the gel–forming, emulsifying, water–binding and fat–retaining capabilities and also to increase the food value and rational use of secondary raw. There exists the opinion that usage of admixtures is only directed at the improvement of the economic indices in finished products. In reality, a series of admixtures improves taste, smell, consistency and commodity appearance of the product and assists in increasing their assimilability.

Many components possessing such properties are known in food industries among which are phosphates, alginates, carragheenins (used as thickeners, stabilizators, structure–formers, emulgators). Albuminous texturators, soy–bean isolates and concentrates are used for improving functional properties and increasing the foodstuffs food value. The excellent emulgators are albuminous preparations produced from soy–beans, peas, lupines, and other beans and cereals. Use of the unique natural milk raw and the products of its processing is being currently an important task confronting the dairy industry which has not been completely solved. Milk albumens are notable for their high nutritive value, high content of the sulphur – containing amino acids, good solubility, high viscosity and water- binding capability. Milk albumems are attributed to the biologically valuable ones owing to the content and ratio of indispensable amino acids. It Is established that use of milk albuminous admixtures at producing foodstuffs improves their organoleptic indices and increases the food value of finished products. The problem of the rational use of separate milk components and milk products on the principles of the wasteless technology is an actual task for food industry. One of the main sources of the products from milk albuminous carbohydrated raw is whey, a part of which is not now used. At present there have been developed various production technologies of drinks and concentrates from whey with intermediate humidity, wasteless technologies of milk sugar manufacture, ecologically pure production technologies of lactose and the technologies for using milk raw components in the foodstuffs formulations. The dairy industry has at its disposal the significant volumes of secondary resources which are not used effectively enough. Nonfat milk (NFM), whey (W), (cheese whey and crude sour milk curd), casein-albuminous concentrate (CAC), lactobal belong to their number. Among products of the secondary milk raw processing the nonfat dry milk (NFDM), is of a significant volume which are being already used for a long time in food industries of many countries in the production of emulgated foodstuffs. In this connection, it is more advisable and effective to involve milk albumens for their use in food purposes as soluble forms of milk-albuminous concentrates which are produced from nonfat dry milk, whey (cheese whey or crude sour milk curd) by removing water, mineral substances, lactose with simultaneous concentrations of albumens. Production of milk coprecipitates, that is the products of joint precipitation of casein and whey albumens, is considered to be perspective. They have valuable functional characteristics and a higher nutritive value, in comparison with casein, that is explained by an increased concentration of sulfur –containing amino acids at the expense of whey albumen fraction.

The purpose of this work was development of the formulational multifunctional compositions with soy–bean concentrate and different types of milk albumen–containing products having high functionally-technological characteristics. Nonfat dry milk, lactobal, dry whey and casein-albuminous complex were used as milk admixture components. Studies on optimization of the compositional formulation have been carried out according to the plan made on the basis of Greek-Latin squares. Realization of the plan has been carried out at five levels. The planning matrix is presented in Table 1. When making the planning matrix, the quantity of soy-bean concentrate (SC) was taken to be equal to 100 % and the content of milk albuminous–carbohydrated components was calculated in percents to the SC total mass. Proceeding from this, the X1 factor was not used while making the matrix. In accordance with action levels of the factors and the experiment planning matrix the formulational compositions have been made and such their main functionally-technological properties as swelling, fat–absorbing and water– absorbing capabilities, degree of penetration have been studied. For the purpose of evaluating the food value of each composition in the "MathCad Professional 2001" program packet the aminoacidic composition of the formulas had been calculated and the albumens content was determined. Estimation and reveal of crude errors and blunders had been carried out by means of the "Statistic V.6" program according to the coefficients of determinations and those of Darbin-Watson's. To optimize the mixture composition, we used the "Statistic Neural Networks V.4" packet.

Table. 1

Test №	X1	X2	X3	X4
1	1	1	1	1
2	2	3	4	5
3	3	5	2	4
4	4	2	5	3
5	5	4	3	2
6	2	2	2	2
7	3	4	5	1
8	4	1	3	5
9	5	3	1	4
10	1	5	4	3
11	3	3	3	3
12	4	5	1	2
13	5	2	4	1
14	1	4	2	5
15	2	1	5	4
16	4	4	4	4
17	5	1	2	3
18	1	3	5	2
19	2	5	3	1
20	3	2	1	5
21	5	5	5	5
22	1	2	3	4
23	2	4	1	3
24	3	1	4	2
25	4	3	2	1

As a result of the data processing in the "Statistic Neural Networks" program the neuron networks achitecture has been developed which allowed to reveal the "weight" of each factor, to determine interfactor interactions and to optimize the composition formulation. The neuron network architecture is presented in Fig.1. In the initial stage of the investigations it was established that presence of dry whey in the mixture reduces swelling, a degree of penetration, water–absorbing (WAC) and fat– absorbing (FAC) capabilities (Figs.2-5). Presence of such components in the compositions as casein-albuminous complex, nonfat dry milk and lactobal promotes the improvement of all functionally-technological characteristics of the food admixture.

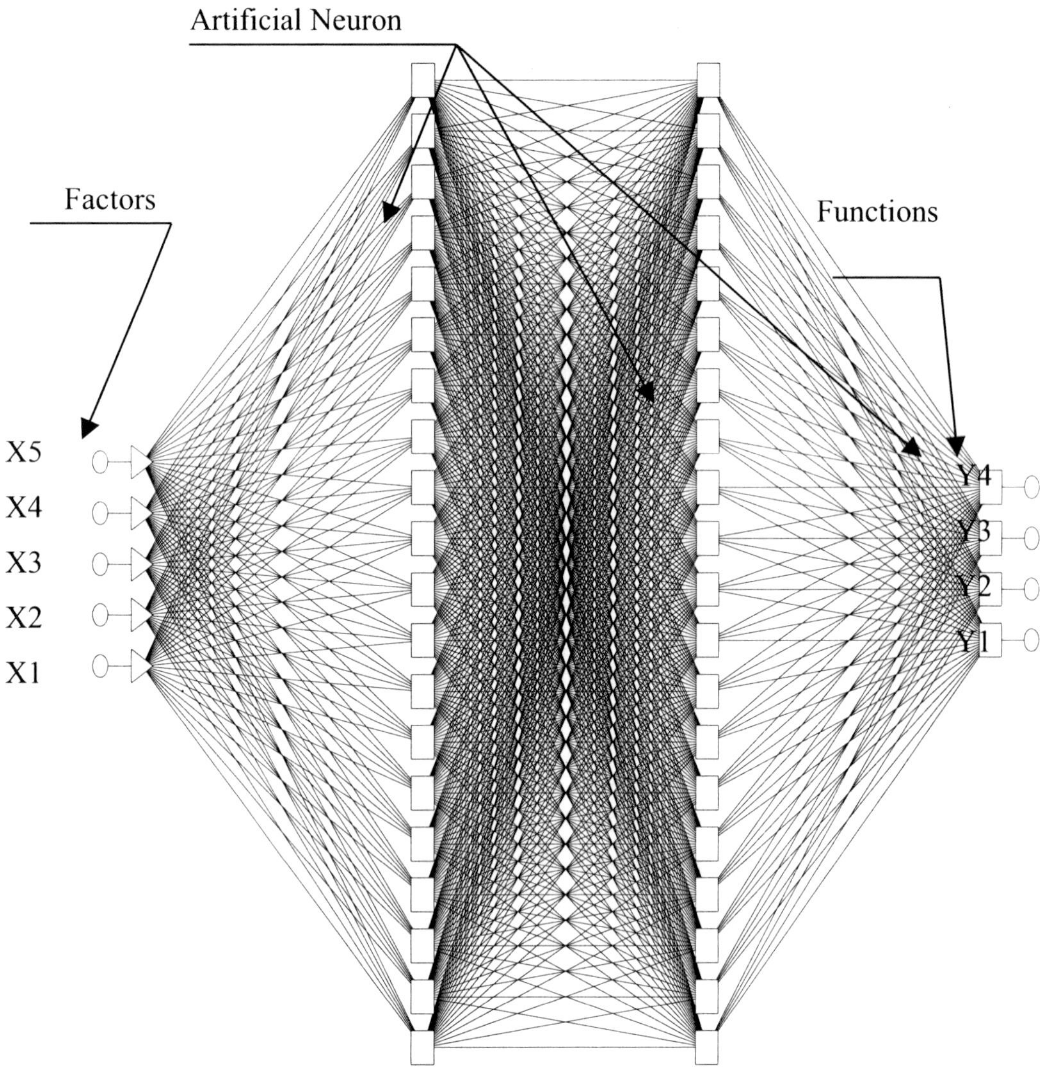
Artificial Neuron
Factors
Functions
X5
X4
X3
X2
X1
Y4
Y3
Y2
Y1

Fig.1

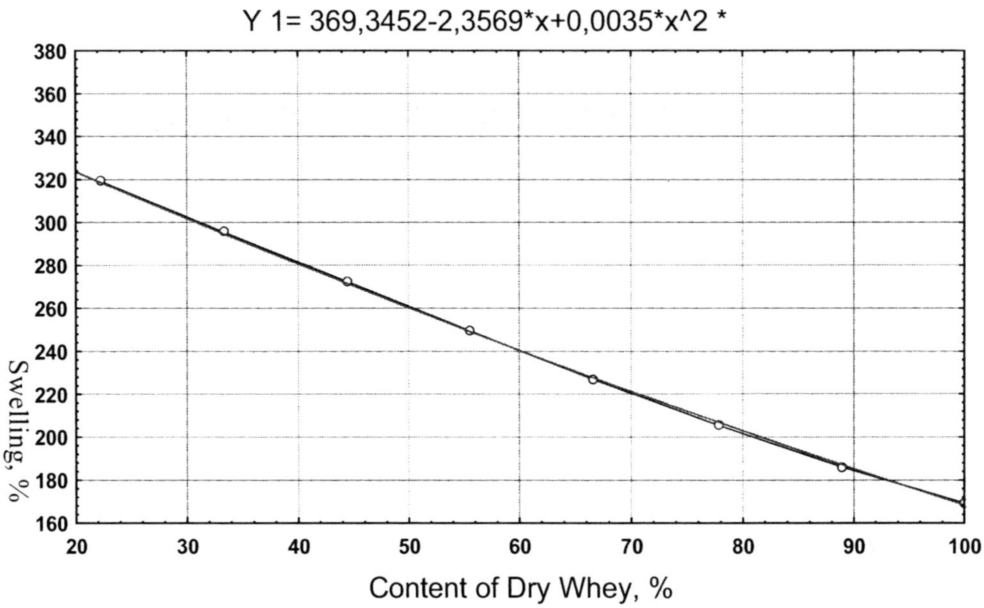

Fig.2

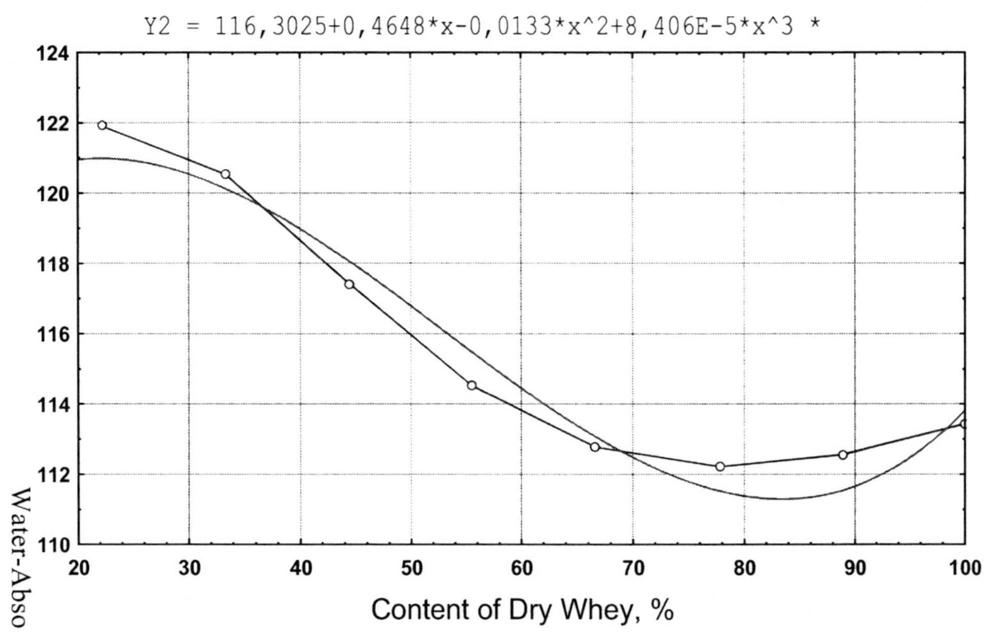

Fig. 3. Water-Absorbing Capability, %

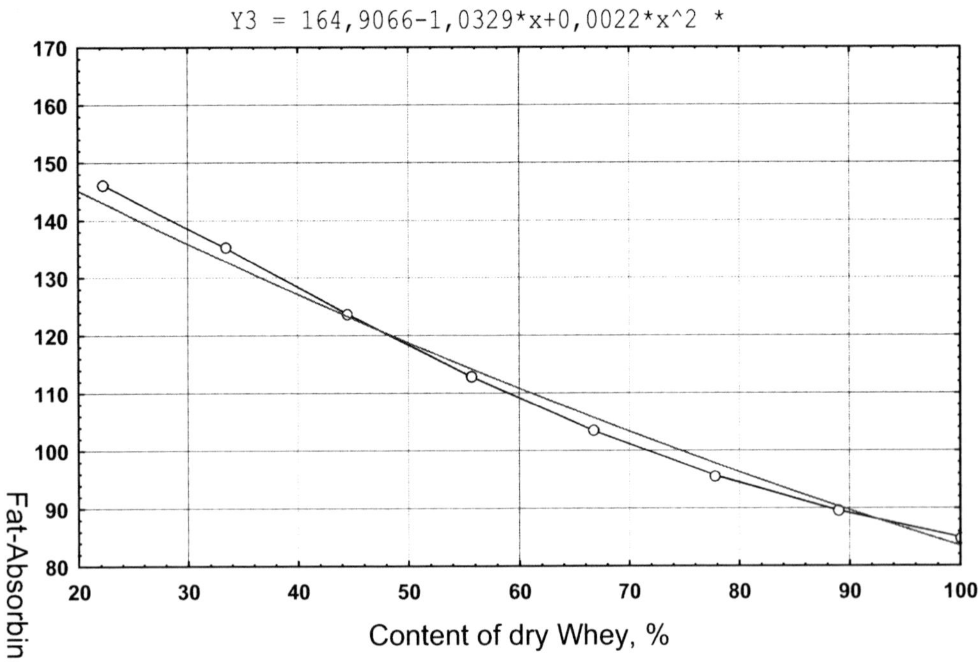

Fig.4

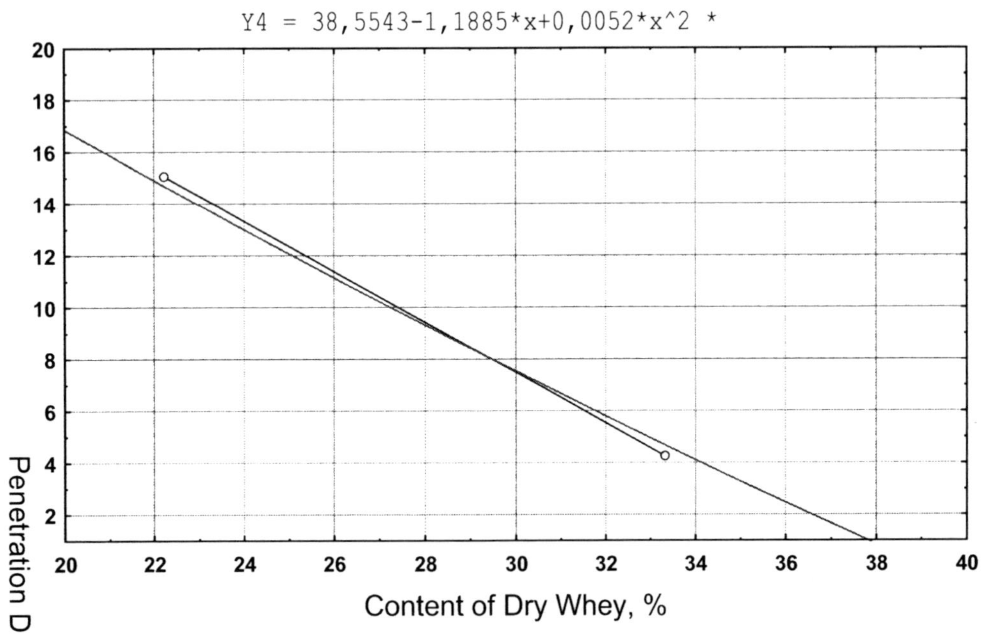

Fig.5

Despite of the positive dynamics of cheese whey on reducting a penetration degree when producing meat products, the most important capabilities are WAC and FAC and swelling. Proceeding from the results obtained, dry cheese whey has not been considered in the subsequent analysis when designing the composition.

Determination of the components quantity ratios has been conducted by means of the "Statistic Neural Networks" packet with consideration of the interfactor interactions. Figure 6 presents the content optimization of nonfat milk and lactobal in the mixture of soy-bean concentrate by swelling of the composition.

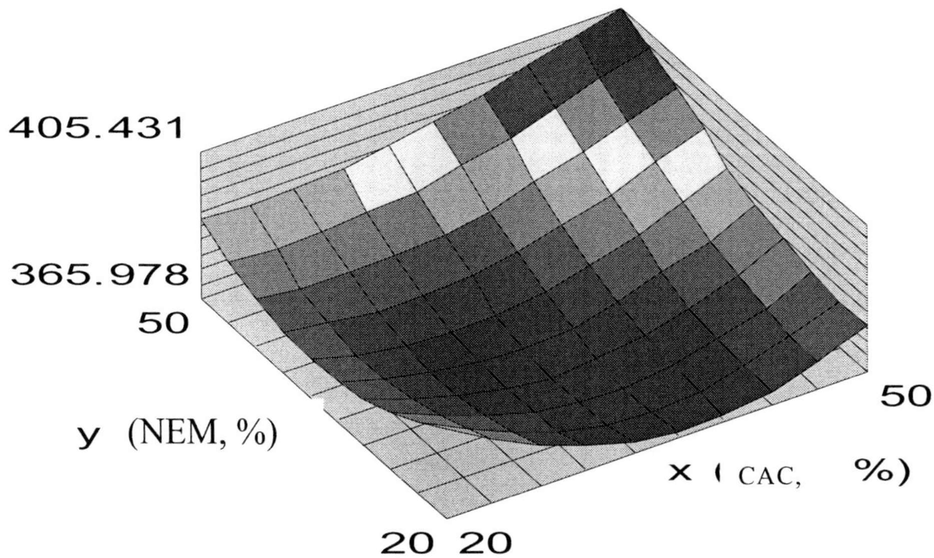

Fig. 6

The following conventional symbols are taken in figures 2-6:

Y1- swelling (in %);
Y2- water–absorbing capability (in %);
Y3- fat–absorbing capability (in %);
Y4- degree of penetration (in %);
X1- content of soy–bean concentrate (in %);
X2- content of nonfat dry milk (in %);
X3- content of lactobal (in %);
X4- content of dry whey in the mixture (in %);
X5- content of casein-albuminous complex (in %).

Optimization of each factor and consideration of all possible variants of interfactor interactions allowed us to reveal six variants of the mixtures having high functionally-technological indices. The functional characteristics of the composition developed by us are presented in Table 2.

Table 2

Composition №	Swelling, %	Water–Absorbing Capability, %	Fat–Absorbing Capability, %	Penetration Degree, %
1	438	141	115	3,2
2	473	145	118	1,7
3	468	142	117	4,4
4	392	137	113	0,4
5	407	135	113	5,3
6	430	136	114	7,2

The optimized formulational compositions are supposed to the used as food admixtures at production of foodstuffs. It is necessary to take into account the technology of the product manufacture and its formulation in each specific case. Design of a food admixture can be carried out both by the specific functionally-technological index and by the complex of properties; so, it is advisable to use a generalized quality criterion.

INDEX

E

F

G

H

I

L

M

N

W